LIGHT MICROSCOPY

The INTRODUCTION TO BIOTECHNIQUES series

Editors:

J.M. Graham MIC Medical Ltd, Merseyside Innovation Centre, 131 Mount Pleasant, Liverpool L3 5TF

D. Billington School of Biomolecular Sciences, Liverpool John Moores University, Byrom Street, Liverpool L3 3AF

CENTRIFUGATION

RADIOISOTOPES

LIGHT MICROSCOPY

ANIMAL CELL CULTURE

GEL ELECTROPHORESIS: PROTEINS

PCR

Forthcoming titles

MICROBIAL CULTURE

ANTIBODY TECHNOLOGY

GENE TECHNOLOGY

LIGHT MICROSCOPY

David J. Rawlins
Department of Cell Biology, John Innes Institute,
Colney Lane, Norwich NR4 7UH, UK

Present address: Biotechnology and Biological
Sciences Research Council, Polaris House,
North Star Avenue, Swindon SN2 1UH, UK

*β*IOS
SCIENTIFIC
PUBLISHERS

In association with the Biochemical Society

©BIOS Scientific Publishers Limited, 1992

First published in the United Kingdom 1992 by
BIOS Scientific Publishers Limited,
St Thomas House, Becket Street, Oxford OX1 1SJ
Tel: +44 (0)1865 726286. Fax: +44 (0)1865 246823.

Reprinted 1995

A CIP catalogue for this book is available from the British Library.

ISBN 1 872748 11 2

Typeset by Enset Photosetting Ltd, Bath, UK
Printed by The Alden Press Ltd, Oxford, UK

Preface

Light microscopy is one of the oldest scientific techniques and also one of the newest. The word 'microscope' was first used by Faber in 1625 [Turner, G.L'E. (1990) *R.M.S. Proc.*, **25,** 423]. On the other hand, equipment for one of the newest light microscopic techniques, confocal fluorescence microscopy, has only become commercially available in the last 5 years or so but already this technique has been extensively used in both the biological and physical sciences. Indeed, there has been a renaissance in the use of all types of light microscopy in the last decade and there is a need for a straightforward, practical guide to the capabilities and basic operation of the light microscope for students of the biological sciences. It is hoped that this book will fill that need.

This book is aimed primarily at those new to light microscopy but it is also hoped, however, that more experienced users will find the information provided here useful for background information on familiar techniques, to correct bad habits or as a source of information about some of the less commonly used techniques. Part 1 starts with a guide to choosing the best type of light microscopy for a particular specimen followed by a brief, simplified description of the theory of microscope optics. The parts of a microscope are then described and there follows a discussion of the principles of the different types of imaging that can be performed with the light microscope: bright field, phase contrast, fluorescence, dark field, polarized light, reflected light and Nomarski differential interference contrast. Part 1 ends with a chapter on three-dimensional microscopy including confocal microscopy.

Part 2, Techniques and Applications, starts with information on the stages of specimen preparation for light microscopy and then covers the practical application of each of the foregoing imaging methods. A section on maintenance of the microscope is also included. Four case studies are then presented which demonstrate the use of several imaging methods to gain the maximum information about a particular specimen. Finally, there are chapters on photomicrography, measurement and video microscopy.

David Rawlins

Acknowledgments

This book arose out of a course that I organized at the John Innes Institute in February 1991. I should especially like to thank David Flanders for helping to run this course and for designing the figures for the sections on Nomarski and polarized light microscopy. I should also like to thank Kim Goodbody who helped in the production of the fibroblast specimens and my other colleagues who lent their own specimens: David Flanders, Luise Janniche, Jan Peart, Coral Robinson and Trude Schwarzacher.

I am grateful to John Graham and David Billington for carefully editing the manuscript. Finally, I should like to thank my wife and colleagues for their support while I was writing this book.

Contents

PART 2: TECHNIQUES AND APPLICATIONS

8. Case Studies   99

9. Measuring Down the Microscope   107

Abbreviations

3D	Three-dimensional
CCD	Charge coupled device
CLSM	Confocal laser scanning microscope
DABCO	Diazabicyclo[2,2,2]octane
DAPI	4′,6-Diamidino-2-phenylindole
DIC	Differential interference contrast (Nomarski)
EM	Electron microscopy
EPM	Epi-polarization microscopy
EPU	Eyepiece unit
FITC	Fluorescein isothiocyanate
ISIT	Intensified SIT
LM	Light microscopy
NA	Numerical aperture
PC	Personal computer
PEG	Polyethylene glycol
PVA	Polyvinyl alcohol
RCM	Reflection contrast microscopy
RGB	Red, green, blue
SIT	Silicon intensifier target
TSM	Tandem scanning microscope
TRITC	Tetramethyl B rhodamine isothiocyanate
UV	Ultraviolet

1 What Sort of Microscopy Should I Use?

How do you choose the sort of microscopy to use for a particular specimen? This question ideally should be answered in the planning stages of an experiment. It is of little value to prepare a beautiful sample if the ideal way of viewing it needs equipment that is not available. However, let us assume that you have a specimen which you want to know how to examine. Look at the top of *Figure 1.1*.

The first question to ask is how small is the specimen? As an example, let us say that you want to examine the different structures inside a small flower. For this, a stereo or dissecting microscope is ideal (*Figure 1.2*). Modern stereo microscopes give very good quality, bright images up to about 64×. They can be operated in transmitted light (light from underneath passes through the specimen before it gets to the lenses) and epi-illumination (light from above is reflected off the surface of the tissue – often from optic fibre, cool light sources) modes. With transmitted light, either bright field (ordinary white light) or dark field (reflective objects e.g. autoradiographic silver grains glow brightly against a black background) can be used. As there are effectively two microscopes mounted next to one another, one to each eye, a stereo or three-dimensional (3D) view of the specimen is produced. Moreover, stereo microscopes are often fitted with a camera. Another way a specimen can be examined if only low magnification is needed is with a hand lens. Hand lenses were used for very many years to look at such specimens. Although the stereo microscope has largely superseded them, hand lenses are still invaluable for examining specimens in the field.

If you want to look at something about the size of a cell or smaller then you will need a compound (proper name for an ordinary) microscope. The next step is to consider how the specimen is mounted. If it is on a Petri dish or in a well of a multiwell plate then you will need an inverted microscope (*Figure 1.3*). This allows the specimen to be viewed from underneath: the microscope has a large space between the stage and the condenser to accommodate the sample. The optics are somewhat compromised using this sort of specimen as the surfaces of plastic Petri dishes are not optically flat or clear like glass. Also, the objectives (the main lenses) are further away from the actual specimen than is ideal. However, for many applications of this sort, they are very good and most of the types of

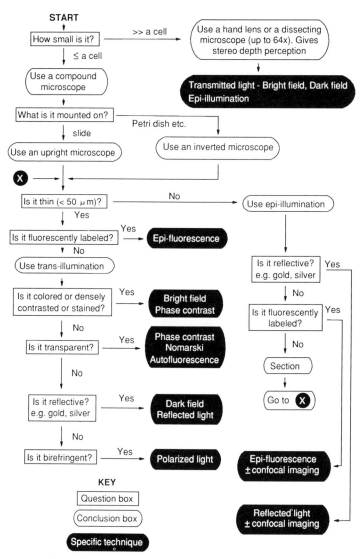

START
How small is it? → >> a cell → Use a hand lens or a dissecting microscope (up to 64x). Gives stereo depth perception
↓ ≤ a cell

Use a compound microscope

Transmitted light - Bright field, Dark field Epi-illumination

What is it mounted on? — Petri dish etc. → Use an inverted microscope
↓ slide
Use an upright microscope

X →

Is it thin (< 50 μm)? — No → Use epi-illumination
↓ Yes

Is it fluorescently labeled? — Yes → **Epi-fluorescence**
▼ No

Use trans-illumination

Is it reflective? e.g. gold, silver — Yes
↓ No

Is it colored or densely contrasted or stained? — Yes → **Bright field Phase contrast**
↓ No

Is it fluorescently labeled? — Yes
↓ No

Is it transparent? — Yes → **Phase contrast Nomarski Autofluorescence**
↓ No

Section

Is it reflective? e.g. gold, silver — Yes → **Dark field Reflected light**
↓ No

Go to **X**

Is it birefringent? — Yes → **Polarized light**

Epi-fluorescence ± confocal imaging

KEY

Question box

Conclusion box

Reflected light ± confocal imaging

Specific technique

"It" = specimen to be visualized

FIGURE 1.1: *What sort of microscopy should I use?*

microscopy described below can be carried out using an inverted microscope. They are also used for microinjection experiments – the microinjection apparatus is mounted on the stage and the specimen viewed from underneath.

If the specimen is mounted on a slide, you should use an upright (ordinary) microscope (*Figure 1.4*). Most specimens are thin enough and transparent enough to permit use of one of the transmitted light methods. However, if the specimen is thicker than about 50 μm and/or rather opaque, you will probably have to use epi-illumination methods. These methods include

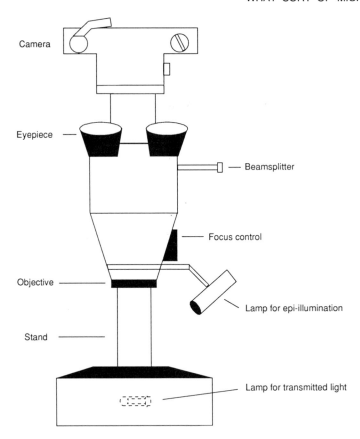

Camera

Eyepiece

Beamsplitter

Focus control

Objective

Lamp for epi-illumination

Stand

Lamp for transmitted light

FIGURE 1.2: *Schematic diagram of a stereo or dissecting microscope.*

reflected light and confocal fluorescence microscopy. Confocal microscopy is an extremely useful technique for visualizing 3D structures labeled with fluorescent molecules.

Fluorescence microscopy is a very powerful technique that allows specific localization with very little background of very small numbers of labeled molecules in the tissue. In combination with confocal microscopy which allows examination deep into thick tissue labeled with fluorescent molecules, it is the most rapidly expanding branch of microscopy at present. Fluorescence microscopy is nowadays always done with epi-illumination.

Using transmitted light illumination, the first way you should try to observe your specimen is with bright field illumination. This relies on color or contrast in the specimen itself. If there is not much of either, bright field will not reveal much detail. However, it is the method that introduces the least number of artifacts into the image and so should always be considered first. If you are using stained or colored specimens, obviously use bright field first. If the specimen has insufficient contrast for

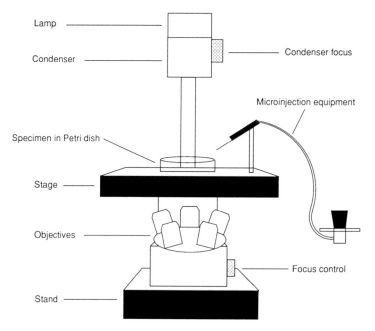

FIGURE 1.3: Schematic diagram of an inverted microscope.

bright field, next try phase contrast microscopy. This uses the ability of biological material to slow down light waves and so produce phase differences between the light from the specimen and that from its surrounding medium. These phase differences cannot normally be seen by the naked eye but with phase contrast microscopy they are converted into contrast that can be seen (see Section 4.2). Phase contrast is also good for stained

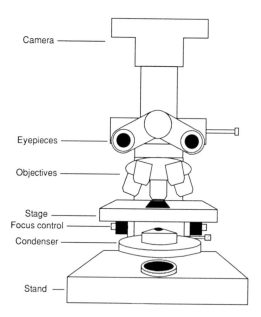

FIGURE 1.4: Schematic diagram of an upright compound microscope.

specimens where it gives more contrast, especially if the staining is rather weak. It is not suitable for heavily stained or thick specimens, however, as the multiple diffractions produced a confused mess.

Nomarski differential interference contrast (DIC) microscopy is an excellent technique for looking at living cells as it produces very high contrast at the edges of biological structures, for example membranes. It also produces very thin optical sections (see Section 4.7) which give 3D information.

For unstained plant tissue where there are cell walls present, or better still lignified tissue or chlorophyll, there is a good chance that you will be able to see something with fluorescence microscopy as these structures will often fluoresce themselves, without any other labeling (autofluorescence). Try using blue light initially or, for chloroplasts, green light.

Dark field illumination is ideal for specimens labeled with silver, gold or other reflective substances where the reflective particles glow brightly against a dark background. The technique is simple and can yield much useful information. Polarized light microscopy is used to study birefringent molecules (e.g. cellulose microfibrils) where different orientations of the molecules are contrasted by darkness or color. Finally, reflected light microscopy has been used most extensively to study sites of adhesion of motile animal cells onto glass surfaces. However, it is also useful for specimens labeled with silver, especially if the tissue is too thick for dark field.

2 Microscope Optics

2.1 Properties of lenses

The lenses which are most relevant for our purposes are converging (convex) lenses. These focus light as shown in *Figure 2.1*. Light from the top of the object (here an arrow) is focused such that the parallel ray passes through the focal point (at a distance f – the focal length) of the lens on the other side. Light passing through the center of the lens is not affected. Where the two rays meet, an image is formed. In this example, the ray from the bottom of the object also passes through the center of the lens and is unaffected. The result is an inverted, magnified image of the arrow.

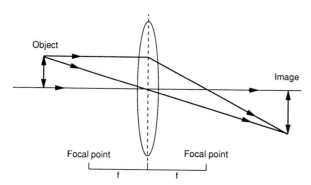

FIGURE 2.1: *Magnification by a simple converging lens. f is the focal length of the lens.*

The way a lens magnifies an object depends upon where it is placed with respect to the focal point of the lens. If it is further away than the focal point, as in the example above, the result is an inverted, magnified, 'real' image. A real image is one that can be photographed or formed on a screen. Two lenses can be combined to produce more magnification using real images as shown in *Figure 2.2*. The result is still a real, magnified image but now it is the right way up. As a practical example of this, when you look through a convex lens, by adjusting the distances between the

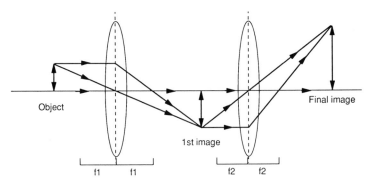

FIGURE 2.2: *Magnification by combined lenses. Two converging lenses combine to give extra magnification. The intermediate and final images are both real.*

object, the lens and your eye, you can get a magnified image. In this case, though, it will *appear* inverted (as the brain inverts everything).

Another way to produce magnification is by creating a 'virtual' image. If the object is placed *inside* the focal point of the lens, the rays do not converge after passing through it, but diverge. An image can still be formed, however, by using an additional converging lens (*Figure 2.3*). In this case the image formed *appears* on the same side of the lens as the object, magnified and the right way up. The image is said to exist 'at infinity' which means it can be seen with a relaxed eye as if viewing a distant object.

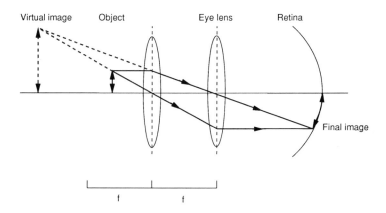

FIGURE 2.3: *Formation of a virtual image. An object placed inside the focal point gives a virtual image which is focused by the eye lens onto the retina.*

A compound microscope can be regarded as two converging lenses (*Figure 2.4*). The first lens produces a magnified, inverted, real image. The second lens produces a 'virtual' image of the first that is focused by the eye to give a greatly magnified image focused at infinity.

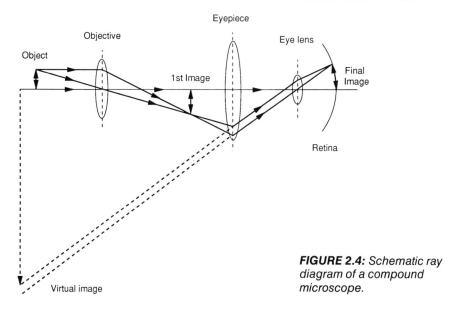

FIGURE 2.4: *Schematic ray diagram of a compound microscope.*

2.2 Resolution, magnification and numerical aperture

Imagine two points in a specimen which are very close together (for example the finest veins in a leaf). If you can see them at all it means light rays emanating from these points are entering your eye (*Figure 2.5a*). The angle α between the light rays from each vein will be small. If it is too small (the light from A–B), both rays will be focused by the eye lens onto the same cell in the retina. In this case, the brain will only see one point and not two. If you look at the leaf more closely (*Figure 2.5b*), you are effectively increasing the angle between the rays. The rays will now be focused on different retinal cells and so the two veins can be distinguished or resolved. What is the smallest distance between two points that can be resolved? If you have normal sight so your minimum distance of focus is about 25 cm, you will be able to resolve points about 150 μm apart. By using lenses to magnify an object, you are effectively increasing the angle of light from two points and improving the resolution beyond that of the eye alone.

Unfortunately, we cannot increase the magnification indefinitely to give any resolution we like. Resolution is critically affected by two things, the light gathering power (numerical aperture or NA) of the lens and the wavelength (λ) of light used. Numerical aperture can be thought of as the size of the cone of light coming from each point in the specimen that can get into the lens (*Figure 2.6*). This is determined by the formula:

$$NA = \eta \times \sin\alpha,$$

where η is the refractive index of the medium between the specimen and

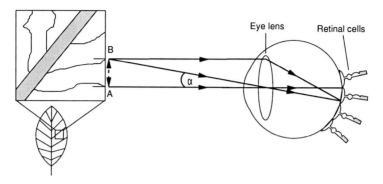

a.

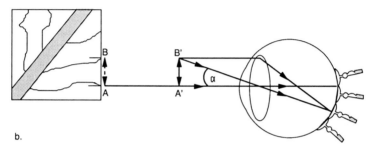

b.

FIGURE 2.5: *Resolution in an image.* **a.** *Light from the two veins of the leaf fall on the same retinal cell and so are not resolved. A–B is the distance between the two veins.* α *is the angle between the light rays emanating from A and B.* **b.** *Focusing closer (A'–B') allows the two veins to be resolved as* α *is increased and light from the two veins falls on separate retinal cells.*

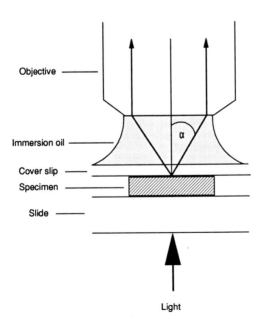

FIGURE 2.6: *Numerical aperture. The size of the angle* α *and the refractive index of the medium between the specimen and the lens determines the numerical aperture.*

the lens and α is the angle between the edge of the cone of illumination and the vertical.

The maximum value sin α can have is 1 so the maximum theoretical NA is the same as the refractive index of the medium between the specimen and the lens – 1.0 for air and 1.515 for an oil immersion lens (see Section 3.1). This theoretical maximum cannot be achieved practically and the highest NA is usually 1.4.

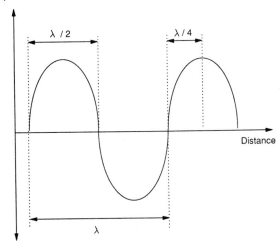

FIGURE 2.7: *Light acts as a wave. The wavelength (λ) is the distance traveled in one cycle, i.e. one peak and one trough.*

Light acts as a wave and so can be thought of as a sine curve with peaks and troughs of amplitude (brightness) at regular time intervals (*Figure 2.7*). The wavelength is the distance the light travels in one complete cycle, that is, one peak and one trough. The relationship between resolution, NA and λ is:

$$\text{resolution} = \frac{0.61 \times \lambda}{\text{NA}}.$$

If we were to use green light with a wavelength of 500 nm and a 1.4 NA lens, the maximum resolution achievable would be:

$$\text{resolution} = \frac{0.61 \times 500}{1.4} = 218 \text{ nm}$$

or 0.22 μm. This is the best resolution (i.e. the closest together two points can be and still be seen as two points) that can normally be achieved with a light microscope, no matter how much the magnification is increased. The much greater resolution achieved by an electron microscope is possible because electrons have a very much smaller wavelength than light.

The magnification needed to increase this distance to that which you can see (i.e. 150 μm at 25 cm) is the maximum *useful* magnification. Greater magnification than this is called *empty* magnification and does not improve resolution. This maximum magnification has been determined as between 500 and 1000 times the NA of the objective depending on the contrast of the points being resolved. This is one of the reasons why you do not see objectives more than ×100 used with ×10 eyepieces as this combination is close to the limit of useful magnification.

a.   b.

FIGURE 2.8: *Effect of wavelength on resolution. Light from two points close together spreads out in a series of waves. **a.** The wavelength is small enough for the two points to be distinguished. **b.** The increased wavelength reduces the resolution.*

The effect wavelength has on resolution is shown in *Figure 2.8*. In *Figure 2.8a*, the peaks (bright areas) and troughs (dark areas) of the waves from the two points are close enough to one another that there is still a trough between the two points. In *Figure 2.8b*, however, the distance between the peaks and troughs (wavelength) is increased: there is now no trough (dark area) between the two points so they cannot be seen as being separate. The way in which waves interact is discussed in more detail in the section on phase contrast microscopy (Section 4.2). For further details on microscope optics see references [1–4].

References

1. Bradbury, S. (1989) *An Introduction to the Optical Microscope*. Royal Microscopical Society Handbook No. 1. Oxford University Press, Oxford.

2. Lacey, A.J. (1989) in *Light Microscopy in Biology: A Practical Approach* (A.J. Lacey, ed.). IRL Press, Oxford, p. 1.

3. Spencer, M. (1982) *Fundamentals of Light Microscopy*. Cambridge University Press, Cambridge.

4. Pluta, M. (1988) *Advanced Light Microscopy, Vol. 1. Principles and Basic Properties*. Elsevier, Amsterdam.

3 Components of a Microscope

Light microscopes are essentially modular and so they all have the same basic parts. However, almost every model of microscope is slightly different from another in the position of the various parts. For this reason I will not describe a particular make of microscope, rather an imaginary, all purpose, upright microscope which represents a hybrid of many different ones. The hybrid microscope contains the following major parts (*Figure 3.1*):

- stand;
- specimen stage;
- objective;
- light source;
- specimen (stage) focus controls, coarse and fine;
- eyepieces;
- condenser;
- field aperture;
- Bertrand lens;
- photo-tube;
- camera.

3.1 Objectives

The objective is the most important part of a microscope – everything else can be regarded as a stand to hold it and the specimen plus a light source to illuminate them. There are several different types of objective but they are usually interchangeable between different microscopes (with the exception of the newer Zeiss objectives). This is because they have the same thread to screw them on and the same tube length (the optical distance between the objective and the eyepiece – usually 160 mm). Zeiss now makes so-called 'infinity-corrected' objectives which do not have a set tube length and are not interchangeable with fixed tube length objectives. Older Zeiss objectives are interchangeable, however.

a.

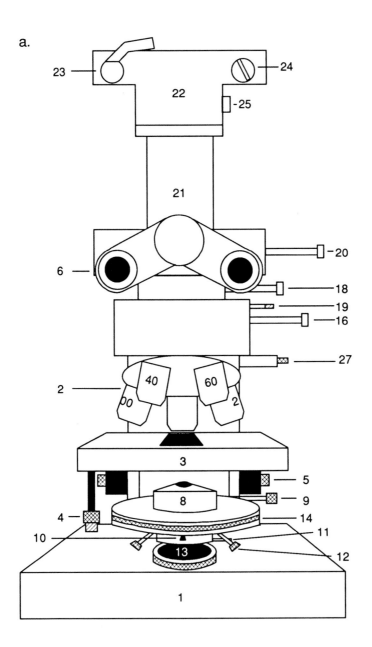

FIGURE 3.1: Diagram showing the parts of a compound microscope. **a.** Front view. **b.** Side view. Key to parts: 1, stand; 2, objectives; 3, stage; 4, stage movement controls; 5, stage focus controls; 6, eyepieces; 7, tungsten light source; 8, condenser; 9, condenser focus; 10, condenser aperture; 11, accessory lens/filter holder; 12, condenser centering controls; 13, field aperture; 14, phase annulus

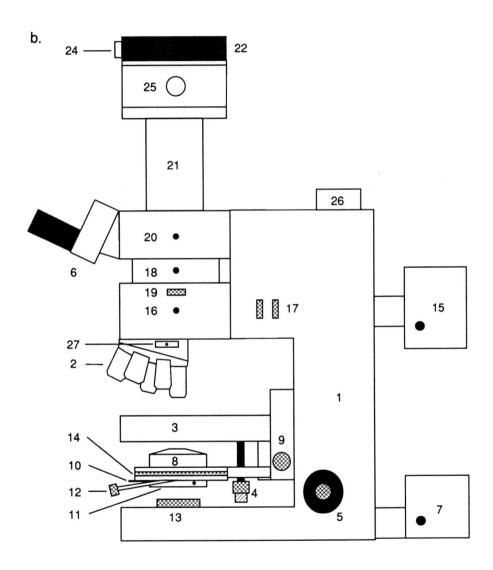

selector; 15, UV light source; 16, fluorescence filter selection control; 17, epi-illumination intensity controls; 18, Bertrand lens; 19, supplementary fluorescence filter; 20, beam splitter; 21, photo tube; 22, camera; 23, manual winder; 24, film rewind; 25, exposure meter; 26, second camera port; 27, Nomarski second beam splitter.

I have said that there are different types of objective. How do you know what type an objective is? On the side of an objective is a set of identification marks which describe the properties of that lens. For example:

Ph 3
Neofluar 40/0.85 Oil
160/0.17

These refer to the following lens properties:

Phase ring
Type of lens Magnification/Numerical aperture Immersion
Tube length/Cover-slip thickness

- Phase rings will be discussed in Section 4.2. Suffice to say at this stage, if there is 'Ph' and a number, there is a phase ring present. The words 'Phaco' or 'Phase' may be found instead. If there are no such markings, there is no phase ring. The presence of a phase ring can be simply confirmed by unscrewing the objective, holding it up to the light and looking through it from the back. If there is a phase ring present, it will look like a dark ring against a bright background.

- Type of lens. The word Neofluar means that the lens will transmit ultraviolet (UV) light. Fluor, Fluotar and UV are other names for ultraviolet transmitting lenses. These lenses are essential for fluorescence microscopy with, for example, the DNA dye 4′,6-diamidino-2-phenylindole (DAPI) which is excited by UV light; while fluorescence using other wavelengths can be done with almost any lens.

The word 'apo' on an objective is an abbreviation for apochromatic and means that the lens is highly corrected for chromatic aberration. Chromatic aberration is the annoying property of a simple lens to focus different wavelengths of light differently. With uncorrected lenses (for example those usually found in children's microscopes) colored fringes will be seen around everything. Apochromatic objectives are highly corrected to avoid this fault. Part of the correction involves the eyepieces, however, so ideally eyepieces and lenses from the same manufacturer should be used. Finally, 'Plan' means that the lens gives a flat field of view, that is, everything is in focus across the whole field. Older non-plan lenses are in focus in the middle of the field but out of focus towards the edges. The best lenses for normal work are therefore Plan–apo lenses.

- Magnification means only the magnification of the objective, not the total magnification. The eyepieces and magnification changer should be taken into account when calculating the total magnification. Better still, calibrate the magnification for yourself (see Section 10.4): 40 means forty times. In this book magnification will be described as '×40' although '40×' is used in some other texts. The second number is the numerical aperture (NA; see Section 2.2). For most purposes, you should use the highest NA you can, as this will give the best resolution and contrast. Some objectives have a movable ring which controls the NA and can be useful for increasing the depth of field of a lens and also in dark field microscopy (see Section 7.4).

● Immersion refers to whether the lens is designed to work with a liquid between the lens and the cover-slip. If the words 'Oil', 'Imm' or 'Oel' are present, then immersion oil should be used. 'W' means water. Some lenses have a movable ring called a correction collar (not the same as the NA adjustment ring mentioned above) which allows the use of different immersion media, for example glycerol or water with or without a cover slip. If there is no indication that the lens is an immersion type, it is called a 'dry' lens. In this case, do *not* use immersion oil or anything else otherwise the result will be a really horrible image (e.g. *Figure 7.8a*).

● Long working distance lenses are designed to focus with the lens a long way from the specimen (up to 1 cm or more in some cases). These are for specimens that, for example, are inside a Petri dish, and so are normally found on inverted microscopes. They will be indicated by the letters 'LWD', 'L' or 'LD' for long- (up to about 5 mm), 'ULWD' for ultra- and 'SLWD' for super-long distance (up to 10 mm or more).

● Tube length is the physical distance between the eyepiece and the objective. It is nearly always 160 mm (except for the newer Zeiss objectives). Cover slip thickness is important as objectives are designed assuming a particular thickness of cover slip. The stated cover slip thickness is always 0.17 mm which corresponds to a '1½' thickness cover slip. The term '160/-' (usually on oil immersion lenses) implies that the cover slip thickness is not important. However, it is still a good idea to use 1½ thickness for everything (see Section 6.2.8).

3.2 Condenser

The function of the condenser is to focus a cone of light of uniform intensity onto the specimen. Like an objective, the condenser has a numerical aperture and can be highly corrected for chromatic aberration (apochromatic). It is mounted below the stage (or above it on an inverted microscope) and is adjustable in all directions. This allows it to be focused and positioned centrally on the optical axis of the microscope. The condenser can have a maximum numerical aperture similar to that of the best objectives (i.e. 1.4). The condenser aperture allows the condenser NA to be matched to the NA of the objective. *Figure 3.2* shows that using a low power objective with a fairly low NA, the NA of the illumination is also set low. For a high power objective with a high NA, the NA of the illumination is set higher. Ideally, the NA of the illumination should be such as to utilize 70–90% of the NA of the objective (*Figure 3.3*). If the NA is set too large, reflections

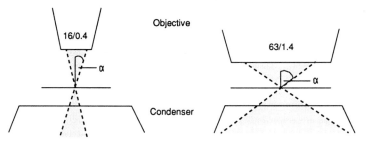

FIGURE 3.2: *Condenser numerical aperture. With low power objectives, a narrower cone of light (smaller NA) is used compared to high power objectives. α is the angle between the edge of the cone of light entering the objective and the vertical.*

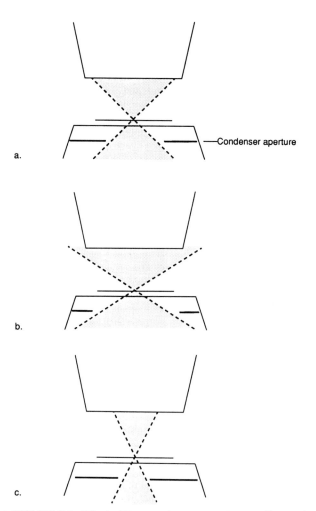

a.

b.

c.

FIGURE 3.3: *Effect of the condenser aperture. a. Correctly set, the illumination fills 70–90% of the front lens of the objective. b. Aperture too far open. Stray light will cause flare and loss of contrast. c. Aperture set too small. Only a very small area of the objective receives light. Resolution is poor and artifacts are caused.*

inside the objective will cause glare and a marked loss of contrast. If, on the other hand, the NA is set too small, only a small part of the objective's NA will be used resulting in poor resolution and contrast artifacts (halos around dark objects).

As well as affecting the resolution in the image, the setting of the condenser aperture also affects the contrast and the depth of field, as shown below.

Condenser aperture position	Resolution	Contrast	Depth of field
Open	Good	Low	Small
Partly closed	Medium	Medium	Medium
Closed	Poor	High	Large

The correct setting of the condenser aperture therefore involves a compromise between these three factors. The effects of condenser aperture position are illustrated by *Figure 7.4a–c*. The correct use of the condenser is explained in Sections 7.1.2 and 7.1.5.

In addition to providing illumination at the correct NA, the condenser must also illuminate a sufficient area of the specimen so that the whole field of view is evenly bright. The focal length of the condenser determines the area of illumination, but at medium to high magnifications it is usually sufficient to cover the whole field of view. However, at low magnification, the use of an accessory lens underneath the main condenser or swinging out the top lens of the condenser may be necessary to illuminate the whole field (see Section 7.1.3).

Many condensers also contain phase annuli which are used in phase contrast microscopy and are selected by rotating a ring on the condenser (see Section 4.2). Some also have prisms for Nomarski (Section 4.7) and a patch stop for dark field microscopy (Section 4.4). Finally, there is often a filter holder to hold colored or polarizing filters.

3.3 Eyepieces

The eyepieces magnify the image formed by the objective and are normally ×10. This magnified image is focused at infinity so that you do not have to focus your eyes up close to see it, simply imagine you are looking into the distance, say at a landscape. The eyepieces also form part of the color (chromatic aberration) correction system that started with the objective. Finally, they allow measuring graticules to be inserted so that they are superimposed on the image. Virtually all modern microscopes are binocular but some older ones are monocular.

3.4 Focus controls

On most microscopes the focus controls move the stage although on some the stage is fixed and the microscope tube, objectives and eyepieces move. The controls are usually mounted concentrically with the larger knob being the coarse focus and the smaller the fine focus. The latter is usually graduated in absolute units (e.g. 2 μm per division). There is normally some adjustment for the tightness of movement of these controls, which varies with different manufacturers.

3.5 Bertrand lens, phase telescope and magnification changer

A Bertrand lens (if fitted) is mounted on the main stand above the objectives and is used for inspecting the back focal plane of the objective (see Section 4.1) in order to set up phase contrast microscopy correctly (see Section 7.2). It is usually inserted into the optical path on a slider or by rotating a wheel, and may be labeled 'Ph'. On some microscopes it is focusable and so can be used to inspect for dust or damage to other intermediate image planes that are otherwise hard to view (see Section 6.3.7). A phase telescope does the same job and looks like an eyepiece but with an extra sliding element that allows it to be focused. There may also be a magnification changer (optovar) at this position giving a range of magnifications from ×1 to ×4.

3.6 Beam splitter

This diverts the light from the specimen either to the eyepieces, or to the camera, or both. It is usually a system of push rods that move prisms back and forth. One position will normally bring a photoscreen into view for use in photomicrography (see Section 10.2).

3.7 Field aperture

This is a variable aperture which is usually on the base underneath the condenser. It is used to vary the area of the specimen that is illuminated and is also used in setting up bright field illumination (see Section 7.1).

3.8 Light source

The word 'lamp' is used here to describe the bulb, its housing and the collector lens. The lamp is usually mounted on the microscope, except in less expensive microscopes where it is external, and a mirror is used to direct the light into the condenser. The bulb is contained in a lamp housing with a collector lens and centering controls. The housing is designed to convect heat away and so should not be covered. Bulbs for transmitted light are either tungsten (like domestic light bulbs) or tungsten/halogen (like projector lamp bulbs). Both are designed to give a particular color temperature or whiteness at a given voltage that matches tungsten rated films very well (see Section 10.3). A ground glass screen is often fitted to even out variations in the intensity of light over the filament.

For fluorescence microscopy, high pressure mercury bulbs are usually used. These emit UV light as well as visible light and are much brighter than other types of bulb. Because the light is much bluer, it matches daylight rated films.

There are three important safety considerations with mercury bulbs:

• First, they have been known to explode. If a bulb explodes, evacuate the room immediately and close the door. Wait at least 1 h for an unventilated room or 30 min if there is air conditioning before going back in to clear up. Treat any liquid mercury spills with a purpose designed spill kit. To minimize the risk of a bulb exploding and to prolong its useful life, a 30 min rule may usefully be adopted, that is, if you switch on, wait 30 min before switching off; if you switch off, wait 30 min before switching on again.

• Secondly, UV light is very harmful to skin and eyes. Do not look at it directly or let it fall on your skin. A cardboard screen can be used to prevent the light that leaks from the lamp housing from reaching the user. If the lamp must be removed from the microscope, wear UV-opaque glasses.

• Finally, the lamps give off ozone which can cause headaches and nausea and may be carcinogenic. If possible, always use mercury lamps in a ventilated room, preferably with air conditioning.

3.9 Photo tube

This is simply a device to mount the camera in such a way that the image is focused on the film. It normally consists of a tube containing an eyepiece or projector lens. This can usually be changed to give a range of magnifications at the film (see Sections 10.4 and 11.4)

3.10 Camera

Most modern microscopes have facilities for taking photographs, normally on 35-mm film. The camera is mounted on the photo-tube in such a way that its plane of focus is the same as that of your eyes when looking through the eyepieces. Photomicrography is described in more detail in Chapter 10.

4 Types of Imaging With the Light Microscope

4.1 Bright field (Köhler) illumination

This is the basic mode of operation of a transmitted light microscope and is quite sufficient for many specimens as long as they contain inherent contrast or color or are stained. The particular method for setting up the microscope for bright field illumination used nowadays was first described by August Köhler near the beginning of this century, and is designed to produce a large area of even illumination from a small light source. It forms the basis of virtually all other types of transmitted light microscopy.

To understand how Köhler illumination works, it is useful to consider two sets of light rays that coincide but do different jobs. One set is called the illuminating rays and the other the image-forming rays (*Figure 4.1*). As you look through the eyepieces, as well as the image formed on your retina, there are also a number of other images formed at intermediate positions along the optical path between the specimen and your eye. Some of these intermediate planes can be examined using the Bertrand lens or phase telescope (see Section 6.3.7).

Let us consider the illuminating rays first; remember that their purpose is to form a wide area of even illumination over the specimen. Light from a point on the bulb filament is focused to a point at the condenser aperture. Therefore, if you were to use a Bertrand lens to view this plane, an image of the filament would be seen at the condenser aperture. Meanwhile, at the field aperture, there is just a wide area of illumination, not a focused image. From the condenser aperture, the light is next focused at a point inside the objective called the back focal plane. Again, this can be examined using a Bertrand lens and an image of the filament will be seen. As before, at the intermediate position, this time the specimen, there is the desired wide area of illumination. The final point at which the filament is focused is on the eye lens which projects the wide area of illumination onto the retina. A point to note is that most manufacturers fit ground glass screens into their lamp housings. This is to compensate for uneven illumination from different points on the filament. Therefore, in practice

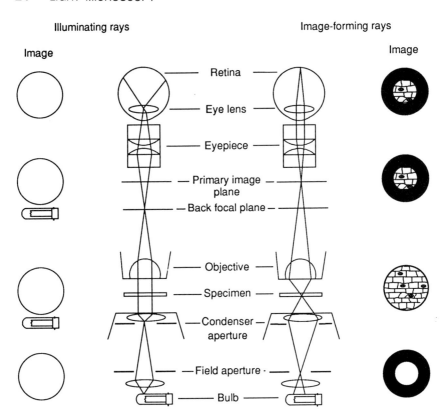

FIGURE 4.1: *Köhler illumination. The illuminating and image-forming rays are shown separately together with the images formed at the intermediate image planes.*

the surface of the ground glass screen is used as the light source and this will be seen at the intermediate image planes. You can see for your-self that the above description is correct by setting up Köhler illumina-tion, removing the ground glass screen and examining the image of the filament at the intermediate image planes with a Bertrand lens or phase telescope.

Now let us consider the image-forming rays. The specimen should now be illuminated by an even field of light. Light coming from the specimen will be focused by the objective at what is called the primary image plane. At this point it is magnified but upside down. The eyepiece lenses turn it the right way up again and form an image a small distance from the top lens of the eyepiece. If you now place your eye so your eye lens is at this point (the exit pupil of the eyepiece), a focused image will appear on the retina. Note too that the field aperture is optically at an equivalent position on the optical path and so is superimposed on the image of the specimen. This is important for setting up bright field illumination (Section 7.1). For more details on Köhler illumination see references [1–3].

4.2 Phase contrast microscopy

Phase contrast microscopy is very useful for specimens that have little inherent contrast under bright field illumination. It works by converting phase differences between light passing through a biological specimen and that passing through the surrounding medium, which the eye cannot normally see, into amplitude (brightness) differences which the eye can see. It does this using the process of interference. To understand phase difference we need to consider light as a wave which has a wavelength, usually denoted by λ (*Figure 2.7*). In *Figure 4.2* there are two waves of different amplitudes (a1 and a2). If the waves are in phase, this means that the peaks of one wave correspond to the peaks of the other and the

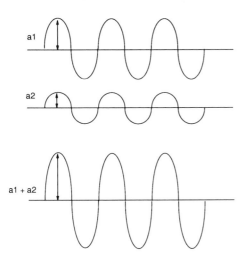

Waves in phase - constructive interference

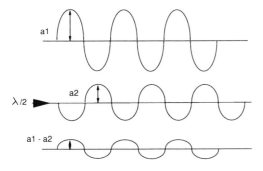

Waves out of phase - destructive interference

FIGURE 4.2: *Interference between light waves in phase and out of phase. a1 and a2 are the amplitudes of the two waves.*

troughs correspond to the troughs. If these waves are combined, constructive interference occurs which results in a wave of the same wavelength as before but the two amplitudes have been added together. If the two waves are not in phase, say the peaks of one wave correspond to the troughs of the other and *vice versa* (half a wavelength out of phase), then destructive interference occurs and the amplitudes are subtracted from one another. These are two extreme cases, in phase and completely out of phase (i.e. by half a wavelength). Constructive interference results in increased amplitude (brightness) and destructive interference in decreased amplitude.

How phase differences produce light and dark (contrast) is shown in *Figures 4.3* and *4.4*. Phase contrast illumination involves placing a ring shaped hole, the phase annulus, in the condenser so that the specimen is illuminated by an annulus of light (*Figure 4.3*). A phase plate is introduced into the objective which is made up of two different types of glass, one in an area equivalent to the phase annulus (the phase ring) and the other everywhere else. The phase plate is designed so that light passing through the portion of the phase plate outside the phase ring is given an additional phase difference of $\lambda/4$ relative to that which passes through the phase ring. *Figure 4.4* shows how this works. Consider two rays coming from the phase annulus in the condenser. They are initially in phase, that is, the phase difference ϕ between them is 0. The left hand ray misses the specimen and passes through the thinner portion of the plate. When light passes through a refractive index greater than that of air it undergoes phase retardation, that is, it is slowed down. Let us call the retardation given to the light ray by the thinner portion of the plate x. The right hand ray passes through the specimen. Now the phase retardation

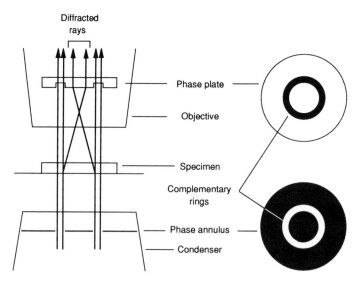

FIGURE 4.3: *The optical elements in a phase contrast illumination system.*

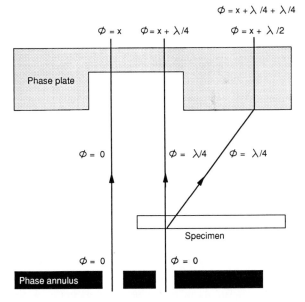

$\phi = x + \lambda/4 + \lambda/4$

$\phi = x$ $\phi = x + \lambda/4$ $\phi = x + \lambda/2$

Phase plate

$\phi = 0$ $\phi = \lambda/4$ $\phi = \lambda/4$

Specimen

$\phi = 0$ $\phi = 0$

Phase annulus

FIGURE 4.4: *Action of the phase plate and annulus. Phase differences (φ) caused by interaction with the specimen are increased if diffracted rays pass through the thick portion of the phase plate. These phase differences are converted into amplitude (brightness) differences by interference.*

produced by biological material is about $\lambda/4$. Some of this light goes straight through the thinner portion of the phase ring where it is retarded by x, so now $\phi = x + \lambda/4$. Some of the light that went through the specimen will also have been diffracted. This will have the same phase difference initially ($\phi = x + \lambda/4$) but will miss the thinner portion of the phase plate and instead will go through the thicker portion of the phase plate. Here it will be retarded by $\lambda/4$ more than the straight-through ray, so that $\phi = x + \lambda/4 + \lambda/4 = x + \lambda/2$. So now we have one ray which has a phase difference of $\lambda/2$ relative to the ray that missed the specimen. From *Figure 4.2* we can see that this phase difference will result in destructive interference between the wave that hits the specimen and the wave that does not. The specimen will therefore appear darker than the background. Variations in the phase retardation across the specimen will give variations in light and dark, thus producing detail in the image. For more details on phase contrast microscopy see references [2–3].

4.3 Fluorescence microscopy

Fluorescence microscopy uses the property of fluorescent molecules called fluorochromes to emit light of a given wavelength when excited by incident light of a different (shorter) wavelength (*Figure 4.5*). By using

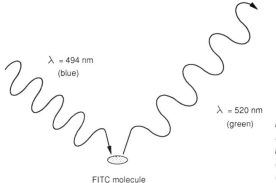

λ = 494 nm
(blue)

λ = 520 nm
(green)

FITC molecule

FIGURE 4.5: *Fluorescence. Blue incident light (λ = 494 nm) excites the fluorochrome (FITC) which emits green light (λ = 520 nm).*

appropriate filters, light of wavelengths other than that being emitted by the fluorochrome can be cut out, giving extremely sensitive and specific detection of, for example, fluorescently labeled antibodies.

High pressure mercury lamps emit light at UV wavelengths (< 400 nm) which are required for many useful dyes like the DNA-specific 4′,6-diamidino-2-phenylindole (DAPI) . They also give off light at many other wavelengths including blue for fluorescein isothiocyanate (FITC) and green for rhodamine (tetramethyl B rhodamine isothiocyanate; TRITC). Fluorescence spectra of these three commonly used fluorochromes are shown in *Figure 4.6*.

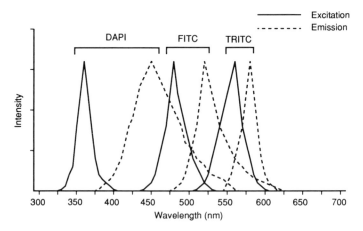

FIGURE 4.6: *Excitation and emission fluorescence spectra of three different fluorochromes: 4′,6-diamidino-2-phenylindole (DAPI), fluorescein isothiocyanate (FITC) and tetramethyl B rhodamine isothiocyanate (TRITC).*

Epi-illumination is used universally nowadays for fluorescence microscopy and *Figure 4.7* shows the principles of this method. For example, for the fluorochrome rhodamine, which is excited by green light and emits red light, there are three different filters (spectrophotometer scans of the transmission of these filters show their abilities to filter different wavelengths—*Figure 4.8a*). The first is called the excitation filter which filters out all wavelengths of light emitted from the mercury lamp except those in the green band (solid line in *Figure 4.7*). The green light then hits a dichroic mirror which has the property of reflecting light at particular wavelengths and transmitting it at others. The mirror reflects green light down through the objective onto the specimen. The fluorochrome emits light at longer (red) wavelengths (represented by the dotted line in *Figure 4.7*). This and reflected incident light go back up through the objective. The dichroic mirror reflects the green incident light as before, but transmits light at the red wavelengths emitted by the fluorochrome. Finally the emitted light is further filtered through an emission filter to make sure that only the light from the fluorochrome reaches the eyepieces.

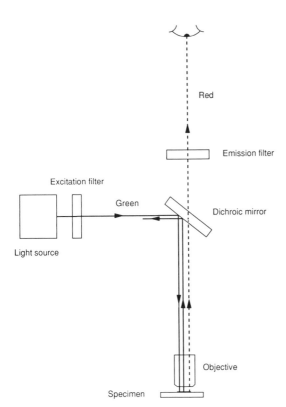

FIGURE 4.7: *A schematic diagram of an epi-fluorescence microscope.*

Transmission spectra for two other commonly used fluorescent filter sets (FITC and DAPI) are shown in *Figure 4.8b* and *c*. Information on the filters in these three sets for the different microscope manufacturers is set out in Appendix C. Examples of the use of epi-fluorescence imaging are shown in the case studies in Chapter 8. For more details on fluorescence microscopy see references [4–5].

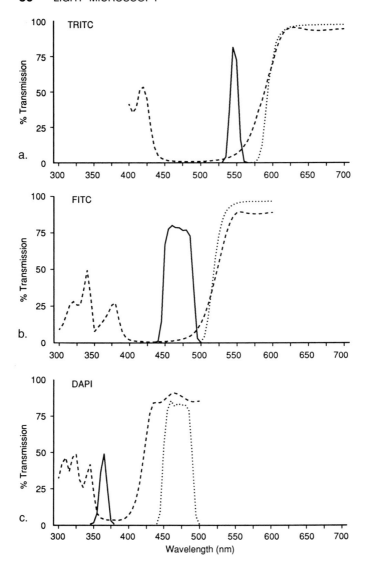

FIGURE 4.8: *Transmission characteristics of the filters used in filter sets for **a.** TRITC, **b.** FITC, and **c.** DAPI. Solid line, excitation filter; dashed line, dichroic mirror; dotted line, emission filter. Zeiss filters were used for these measurements.*

4.4 Dark field microscopy

Dark field microscopy is ideal for autoradiographs or other silver or reflectively labeled specimens (for examples see Chapter 8). Dark field also works well for delineating plant cell walls on unlabeled tissue sections. It shows such reflective structures as bright objects on a dark background. It uses transmitted light and can be performed across a whole range of magnifications from a stereo microscope at ×10 to an upright microscope at ×1000. *Figure 4.9* shows how in dark field, oblique illumination is used, thus with no specimen present all the light from the condenser misses the objective entirely, giving a dark background. However, if a specimen containing reflective structures is placed into the path of this illumination, light that hits such a structure will be reflected at all angles. Some of this light will now reach the objective and so will appear bright.

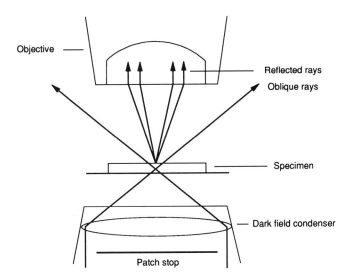

FIGURE 4.9: *Dark field illumination using a patch stop.*

There are two main methods of producing oblique illumination. The first is by blocking out the central portion of the illumination from the condenser with a patch stop (an opaque disk) leaving only oblique illumination (*Figure 4.9*). Condensers containing annuli for phase contrast often have a dark field patch stop fitted as well. Specialized condensers of this type are also available. Patch stop condensers should be oiled to the slide (see Section 7.1.4) for best results. The second method uses reflective surfaces on the sides of a special condenser to produce the oblique illumination. There are two types, one for use with immersion oil between the condenser and the slide (*Figure 4.10a*) and the other for lower magnifications without oil (*Figure 4.10b*). For more details on dark field microscopy see reference [6].

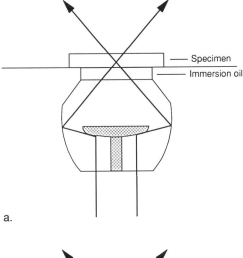

a.

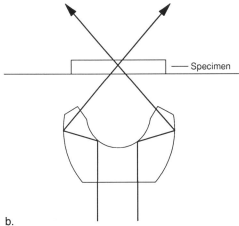

b.

FIGURE 4.10: *Two types of dark field condensers using reflection.* **a.** *Used with immersion oil.* **b.** *Used without immersion oil.*

4.5 Polarized light microscopy

Normal *un*polarized light (i.e. normal light), for example that reaching your eye from *Figure 4.11a*, can be thought of as many sine waves travelling out from the surface of the paper. These sine waves are each oscillating at any one of an infinite number of orientations (planes) around the central axis. In *Figure 4.11a* some of these planes are represented by an arrow. Plane polarized light, produced by a polar, only oscillates in one plane (*Figure 4.11b*) because the polar only transmits light in that plane. If two polars are placed on top of one another, held up to the light and rotated relative to one another, there will be one position where the two transmitted planes coincide, which will appear bright. At 90° to this orientation, where the two transmitted planes are at right angles to one

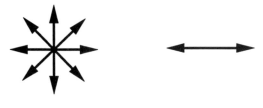

a. Unpolarized   b. Plane polarized

FIGURE 4.11: Polarized light.
a. Unpolarized b. Plane
polarized light.

another, all the light will be stopped. This dark position is called 'extinction' and the polars are then said to be 'crossed'.

Polarized light microscopy uses plane polarized light to analyze structures that are birefringent. Birefringent structures (e.g. a cellulose microfibril) have two different refractive indices at right angles to one another (*Figure 4.12a*). Polarized light microscopy can be used to measure the amount of retardation that occurs in each direction and so give information about the molecular structure of the birefringent object.

If plane polarized light hits a birefringent structure, it will effectively be split into two 'beamlets' which will then oscillate in planes parallel to the two directions of refractive index (*Figure 4.12a*). Now light passing through a substance with a greater refractive index than air is slowed down or retarded (see Section 4.2). Because there are two refractive indices, the two beamlets will be retarded to different extents (*Figure 4.12b*). The beamlet oscillating parallel to the direction of higher refractive index will be retarded more than the other. This means that there will now be a

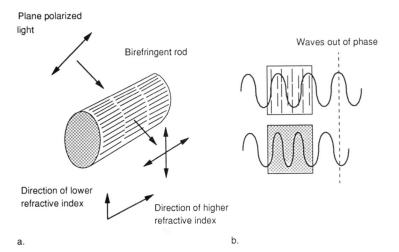

a.   b.

FIGURE 4.12: The effect of birefringence. Plane polarized light hitting a birefringent object is split into two beamlets parallel to each of the directions of refractive index. This results in different phase retardation for each set of beamlets.

phase difference between the two beamlets. We saw with phase contrast microscopy how phase differences lead to destructive and constructive interference. However, in this instance the two beamlets are not in the same plane but are oscillating in planes at right angles to one another. The effect of recombining these two beamlets is to produce a helical or corkscrew-like beam. The shape of the helix depends on the relative phase difference between the two beamlets (*Figure 4.13*). If there is $\lambda/4$ retardation, the helix ends up circular in profile. However, as the phase difference decreases or increases towards 0 or $\lambda/2$ respectively; the helix flattens until at the two extremes it is a straight line.

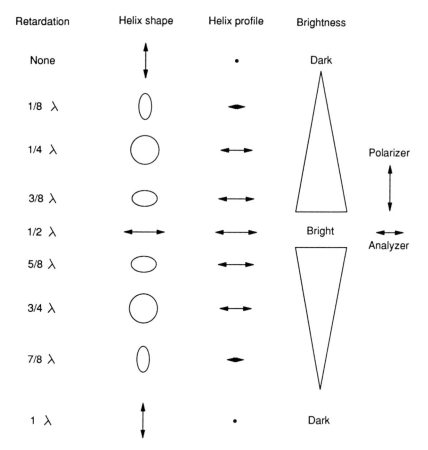

FIGURE 4.13: *The effects of retardation on brightness. Beamlets recombine to form helices which have different shapes depending on the phase retardation. The profile of the helix with respect to the analyzer gives different levels of brightness.*

In order to actually see the effect a birefringent structure has on plane polarized light, we can put another polar behind it which is crossed relative to the first. A diagram of this arrangement, that of a polarized light microscope, is shown in *Figure 4.14*. The first polar is usually called the polarizer and the second the analyzer. Now consider the helices produced by the birefringent structure. If there is no retardation and the helix is flat, the light will effectively oscillate in the same plane as it started out (vertical in *Figure 4.13*). As its profile with respect to the crossed polar or analyzer (horizontal in *Figure 4.13*) is zero, no light will be transmitted and so the image will be dark. As the helix fills out, it will have some component in the same plane as the crossed polar and so will begin to be transmitted. Increasing the phase difference between the two beamlets further will give more and more component in the horizontal plane and so the image will appear brighter and brighter until at λ/2, it will be maximally bright. Thereafter, the reverse situation happens and the image gets darker and darker as the vertical component increases.

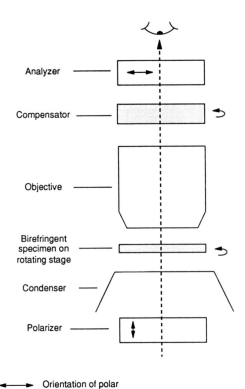

FIGURE 4.14: *The components of a polarizing microscope.*

How can these effects be used? If we now place a second birefringent structure of known properties (called a compensator) in the light path (*Figure 4.14*), we can use it to find the birefringence of the first by rotating one with respect to the other. The relative angles between the two at positions which give maximal brightness and darkness can be used to calculate the birefringent properties of the unknown structure. This in turn gives information about its molecular structure. Polarized microscopy is used extensively in materials science, where sophisticated measuring techniques have been developed to study the physical properties of crystalline structures. A biological example of the use of polarized light microscopy is shown in *Figure 7.15*. For more details on polarized light microscopy see references [6–8].

4.6 Reflected light microscopy (epi-polarization and reflection contrast microscopy)

Epi-illumination using polarized light can be used to visualize specimens that contain reflective structures in them or that produce reflective surfaces by being in very close contact with their support [9–12]. There are two types of reflected light microscopy: epi-polarization microscopy (EPM) and reflection contrast microscopy (RCM). EPM is the simplest of the two and is useful for looking at, for example, silver enhanced gold particles or autoradiographic silver grains, and gives a result similar to dark field. It can also be used with confocal microscopy (see Section 5.2). Epi-polarization microscopy is performed in a manner similar to epi-fluorescence except that instead of a dichroic mirror, there is a semi-silvered mirror which reflects 50% of the light hitting it and transmits the rest (a 50:50

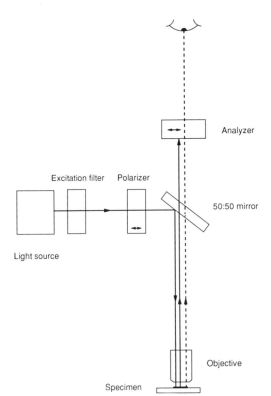

FIGURE 4.15: *The components of an epi-polarization microscope.*

mirror; *Figure 4.15*). Plane polarized light is used to illuminate the specimen so there is a rotatable polar between the light source and the 50:50 mirror. Certain types of reflective surfaces (e.g. silver grains) change the plane of polarization of light slightly after reflection. This reflected light (dotted line) will now be able to pass through the second polar (analyzer). However, light that just reflects off the background (solid line) will not have had its plane of polarization altered with respect to the illuminating light so it will not get through the second polar. What is seen, therefore, is bright silver grains against a dark background.

Reflection contrast microscopy has been most often used for looking at how motile animal cells adhere to a flat surface to enable them to move across it [11], but is also suitable for the same specimens as used with EPM [12]. In this case, reflected light illumination is usually set up on an inverted microscope. It gives a bright image of the so-called focal adhesion plaques against a dark background. These are then usually printed as a negative image so that they look black. RCM uses a similar setup to EPM but has two additional elements. The first is a rotatable quarter wave plate (a device which retards light by a quarter of a wavelength) mounted in the objective just behind the front lens. This produces light that is polarized not in one plane but in a helix – circularly polarized light. This has the property that on reflection, light will pass back through the quarter wave plate but will become plane polarized at 90° to the incident light and will thus pass through the analyzer. In theory, only light that is reflected will get back to the eyepiece. The main advantage of this element is that it cuts down internal reflections in the objective that degrade the image. The second additional element is a condenser annulus similar to that used in phase contrast microscopy. This produces ring-shaped illumination which also helps cut out reflections. EPM and RCM are discussed in detail in references [9–12].

4.7 Nomarski microscopy

Nomarski differential interference contrast (DIC) microscopy is one of the most aesthetically pleasing forms of microscopy. The images that are produced appear three-dimensional because one side of the specimen appears lighter than the other as if light was falling on it and casting shadows. This is in fact an artifact but it is very useful nonetheless. Another useful property of Nomarski illumination is that it produces very thin optical sections (see Section 5.1) which are useful for thick specimens. It is also immune from the light scattering problems that this sort of specimen causes with phase contrast microscopy.

Nomarski is superb for investigating living cells as it is non-invasive and the real-time optical sectioning properties allow the movement of tiny organelles to be followed with ease. It uses polarized light in an ingenious way to produce the 'shadowing' effect. Examples of Nomarski are shown in Section 7.5 and Chapter 8.

To explain Nomarski microscopy let us consider a microscope set up for polarized light microscopy with the polars (polarizer and analyzer) crossed (see Section 4.5) and with the addition of two special beam splitters (*Figure 4.16*). The first is placed in the path of the incident, plane polarized light. This splits the light into two beamlets but without any phase difference between them (*Figure 4.17*). Some of the light passes straight through the slide (the pair of beamlets on the left), misses the specimen and then passes through a second beam splitter. This time, a phase difference of λ/4 is added to the *right hand beamlet only*. We started with no phase difference and now have a phase difference of λ/4. Referring back to *Figure 4.13*, this will give an intermediate brightness once the light goes through the second polar (analyzer), in other words, gray. With the next set of beamlets in *Figure 4.17*, the right hand one *only* hits the specimen. The specimen retards the beamlet by about λ/4 so there is now that phase difference between the two beamlets. The second beam splitter increases this by adding λ/4, giving a total of λ/2. This goes straight through the

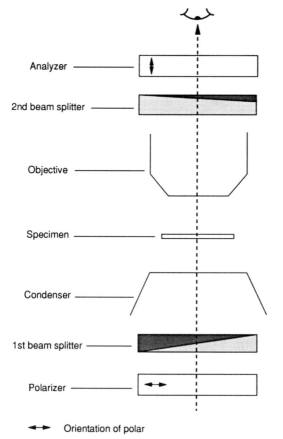

Analyzer

2nd beam splitter

Objective

Specimen

Condenser

1st beam splitter

Polarizer

Orientation of polar

FIGURE 4.16: *The components of a Nomarski illumination system.*

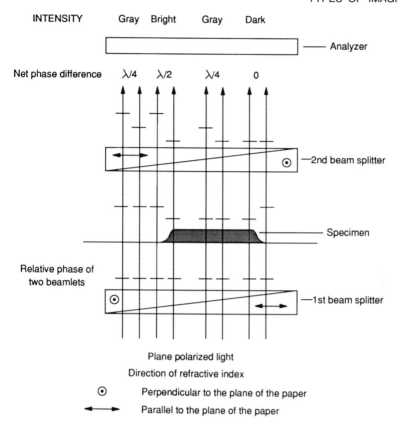

FIGURE 4.17: *How contrast is formed under Nomarski illumination.*

second polar giving a bright image of the edge of the specimen. The third set of beamlets are both retarded by the specimen and then $\lambda/4$ is added to the right hand one. This again gives a gray image. Now with the last set, the left hand beamlet only hits the edge of the specimen and is retarded. The second beam splitter gives a $\lambda/4$ retardation to the right hand beamlet, resulting in an overall phase difference of 0 between the two beamlets. Referring back to *Figure 4.13*, this gives a dark image on this edge. The position of the second beam splitter is variable, allowing the contrast in the image to be varied (see Section 7.5.2). Now we can interpret Nomarski images knowing how they are created. The background is gray, the left hand edge is bright, the center gray again and the right hand edge dark. In practice, the second beam splitter is usually mounted at 45° to the orientation of the image in the eyepieces, so that the shadowing effect appears to come from above and to one side.

It is clear from this explanation that Nomarski illumination picks out edges (or abrupt changes in refractive index) especially. This is very useful in biological research as biological material is full of edges – membranes of all kinds, nuclei, organelles, cell walls, etc. A word of warning, however, should be noted here. Because the images are so pleasing, it is easy to be deceived into thinking something is what it is not! See Section 7.5.3 and reference [13] for further discussion of image interpretation in Nomarski microscopy.

A technique called video-enhanced contrast microscopy [14] deserves a mention here. It uses (generally) Nomarski microscopy with the second beam splitter adjusted to produce very little contrast. Using a very bright light source and image processing, significantly smaller structures can be visualized than with the naked eye. For example, structures as small as 10–50 nm have been visualized including individual microtubules. The resolution is also improved by about a factor of two. It is a real-time technique and has been used to look at the migration of organelles such as endocytotic vesicles along microtubules.

4.8 Other techniques

- Hoffman modulation contrast [15]. This technique gives results very much like Nomarski. It also uses polarized light but instead of two beam splitters it uses a combination of a variable aperture slit and an optical element with bands of different opacities to light. Its main advantage appears to be the cheapness of the components required over those for Nomarski.
- Rheinberg illumination [16]. This is a modification of dark field microscopy whereby the opaque patch stop in the condenser is replaced with a piece of colored filter. It can give very attractive pseudocolored images and is especially good with specimens containing structures oriented in particular directions.

References

1. Bradbury, S. (1989) *An Introduction to the Optical Microscope*, Royal Microscopical Society Handbook No. 1. Oxford University Press, Oxford.

2. Lacey, A.J. (1989) in *Light Microscopy in Biology: A Practical Approach* (A.J. Lacey, ed.). IRL Press, Oxford, p. 1.

3. Spencer, M. (1982) *Fundamentals of Light Microscopy*. Cambridge University Press, Cambridge.

4. Ploem, J.S. and Tanke, H.J. (1987) *Introduction to Fluorescence Microscopy*. Royal Microscopical Society Handbook No. 10. Oxford University Press, Oxford.

5. Taylor, D.L. and Salmon, E.D. (1989) *Meth. Cell Biol.*, **29**, 208.

6. Lacey, A.J. (1989) in *Light Microscopy in Biology: A Practical Approach* (A.J. Lacey, ed.). IRL Press, Oxford, p. 25.

7. McCrone, W.C., McCrone, L.B. and Delly, J.G. (1978) *Polarized Light Microscopy*. Ann Arbor Science, Ann Arbor, MI.

8. Patzelt, W.J. (1974) *Polarized Light Microscopy. Principles, Instruments, Applications*. Leica UK Ltd, Milton Keynes.

9. Verschueren, H. (1985) *J. Cell Sci.*, **75**, 279.

10. Cornelese-ten Velde, I., Bonnet, J., Tanke, H.J. and Ploem, J.S. (1990) *J. Micros.*, **159**, 1.

11. Ploem, J.S. (1975) in *Monunuclear Phagocytes in Immunity, Infection and Pathology* (R. van Furth, ed.). Blackwell Scientific Publications, Oxford, p. 405.

12. Hoefsmit, E.C.M., Korn, C., Blijleven, N. and Ploem, J.S. (1986) *J. Micros.*, **146**, 161.

13. Padawer, J. (1968) *J. R. Micro Soc.*, **88**, 305.

14. Allen, R.D. (1985) *Ann. Rev. Biophys. Chem.*, **14**, 265.

15. Hoffman, R. (1977) *J. Micros.*, **110**, 205.

16. Syred, A. (1990) *Micros. Anal.*, **September,** 21.

5 Three-dimensional (3D) Microscopy

In many cases, biological specimens are thin enough to be considered approximately two-dimensional and so require only a small amount of focusing to see everything there is to see. Where one is interested in the inter-relationships of cellular structures in three dimensions, however, 3D microscopy comes into its own. There are two ways to use a microscope to get information about a 3D structure. The first is to physically cut the specimen into sections, take conventional pictures of each one and print the negatives, trace round important features onto clear plastic film or glass sheets to produce a set of profiles and then stack these on top of one another. The stack can then be photographed from above and, by taking pictures at different angles, the 3D relationships between the structures traced can be determined. Instead of taking photographs onto film, the images can be digitized using a video camera and a frame-store (see Chapter 11) and the process of tracing the outlines and making the 3D reconstructions done by computer. Either way this process is called serial section reconstruction or tomography and this approach has been extensively used in electron microscopy.

The second and more recently developed way of obtaining such 3D information is called optical section reconstruction.

5.1 Optical sectioning

Optical sectioning is analogous to the physical sectioning described above but uses a convenient property of a high power microscope objective to do the sectioning for you without getting anywhere near a knife! This is a great advantage because it means that living tissue or tissue which has been fixed but not embedded in wax or resin (as is necessary for most physical sectioning procedures) can be examined. The way it works is as follows: because high NA lenses have a small depth of field, one can focus down through the tissue at regular spacings along the z axis (Δz) and at each focus level (or focal plane) collect an image, usually with a video or CCD camera, to give a stack of optical sections (*Figure 5.1*). A thin portion of each image will be in focus at any one time. However, there will also be

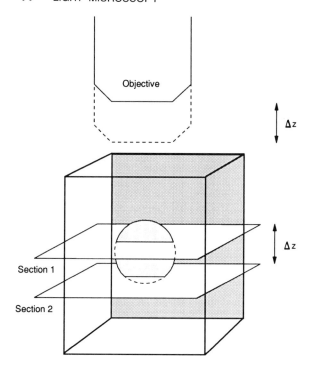

FIGURE 5.1: *Optical sectioning. Two optical sections are formed by focusing up through the specimen by a distance Δz.*

a lot of out-of-focus blur from above and below the plane of focus. This must be removed since it seriously degrades the in-focus part of the image. One approach is to digitize the image and using a computer, calculate what the blurred component is and then subtract it from the total image (deblurring). Another way is to use confocal microscopy, which deblurs the image in real time.

5.2 Confocal microscopy

A confocal microscope allows *in-focus* pictures to be obtained from throughout the depth of a fairly thick (up to about 100 μm) specimen [1–3]. It is most commonly used for fluorescently labeled specimens and often consists of a standard epi-fluorescence microscope with an extra module to do the confocal imaging.

All confocal microscopes work by scanning a small point of light, focused through the objective at a particular depth, over the specimen. Light emitted from fluorescent molecules in the specimen returns through the objective and is filtered out from the shorter wavelength incident light in the same way as for standard epi-fluorescence (Section 4.3). However, before the light reaches the detector (usually a photomultiplier tube – like a video camera) it passes through a small aperture (confocal aperture) that is placed at optically the same point as the focus plane of the scanned beam of light. This has the effect that only light returning from the specimen that is in focus will get through the aperture to the photomultiplier tube. This is shown in *Figure 5.2*. By scanning very quickly, an image can be built up like that on a TV screen.

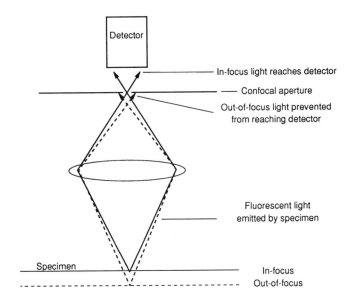

FIGURE 5.2: *Principles of confocal microsocopy. A confocal microscope uses a confocal aperture to remove out-of-focus light.*

The ability of the confocal microscope to remove out-of-focus blur helps to sharpen the image of any specimen that is not extremely thin, and indeed confocal microscopes are often used just for this purpose. It also turns out that the resolution in the plane of focus (in the x and y directions) is improved over an ordinary microscope. So coupled with the brightness and contrast controls that are part of the photomultiplier tube electronics, the confocal microscope is an excellent instrument for looking at any fluorescent specimen.

Most confocal microscopes can also be used for many types of transmitted light microscopy where the electronic control over the appearance of the image can produce greatly enhanced contrast. An optic fiber relay system is placed below the condenser and the image (which is produced in the opposite direction to normal, i.e. light passes firstly through the objective and lastly through the condenser) transferred to the detector. Generally, the image is not confocal.

5.3 Types of confocal microscope

There are two main types of confocal microscope available at the moment. The first type is called a confocal laser scanning microscope (CLSM). Laser light is reflected from a dichroic mirror onto rotating mirrors which scan a very small spot of light over the specimen in the x and y directions parallel to the stage, in a pattern corresponding to the rows of dots (pixels) on a TV screen. The light given off by the fluorochrome in the specimen passes back up through the objective, hits the mirrors again and passes through the dichroic mirror into the detector via the confocal aperture (*Figure 5.3*). Laser light is used as the technique requires a very small, focused spot of high intensity light and the scanning is computer controlled. Several manufacturers sell versions of this type of confocal microscope. They differ mainly in the precise way the scanning is achieved and in the sophistication of the image processing software supplied with the computer.

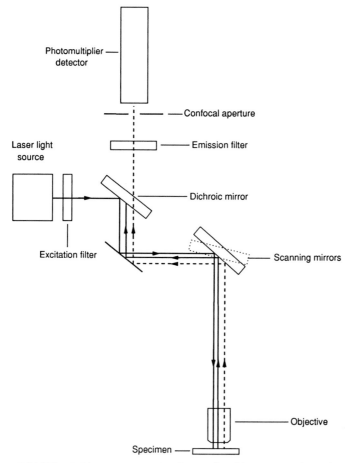

FIGURE 5.3: *The components of a confocal laser scanning microscope (CLSM).*

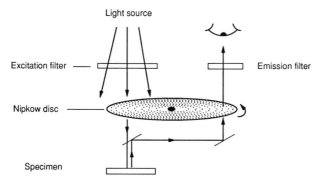

FIGURE 5.4: *The components of the tandem scanning microscope (TSM). The TSM scans a point of light across the specimen using a rotating disk (the Nipkow disk) containing a pattern of many small holes. Emitted light is de-scanned through holes on the other side of the disk.*

The second type of confocal microscope is based on a system called tandem scanning microscopy (TSM). This was the first system to be invented and uses a mercury light source together with a spinning disk containing a pattern of small pinholes (the Nipkow disk; *Figure 5.4*). The holes are arranged in such a way that as the disk turns, each part of the specimen is scanned by a point of light. Light returning from each scanned point either passes through the same pinhole or, in a different system, a complementary one on the other side of the disk. These pinholes act as confocal apertures and produce a confocal image at the eyepieces or relayed to a video camera. The main advantage of this system is that the scanning speed is higher than most CLSMs so that real-time confocal imaging is possible. Also, the color of the illuminating light does not limit the fluorochromes that can be used, as is the case with lasers. However, there are technical problems in filtering the light returning from the specimen from the incident light reflected from the disk itself which limit the device to highly reflective structures. For these, TSMs have produced excellent results [4].

5.4 Optical sectioning with a conventional microscope

Non-confocal optical sectioning is still a useful tool when coupled with computer-based deblurring programs. The price of computers is now low enough that the processing power needed for the deblurring calculations is now not much more than that supplied with a standard PC, and several companies specializing in image processing now supply deblurring programs. For more information about the methods for deblurring conventional optical sections see references [5–7].

The two main advantages of non-confocal optical sectioning are firstly that because all the light emitted from the specimen is collected into the camera instead of being filtered through a small aperture, the signal/noise ratio (i.e. background noise) is often considerably less than with confocal microscopes, especially when they are used with the minimum sized confocal aperture. Secondly, as with the TSM, because a mercury lamp is used instead of a laser, one is not limited to the particular emission wavelengths of the laser for excitation. The lasers that come as standard with most confocal microscopes emit light at two or three different wavelengths, which may not be optimal for a particular fluorochrome. For a more detailed discussion of the advantages and disadvantages of confocal versus non-confocal optical sectioning see reference [7].

5.5 Three-dimensional reconstruction

It is all very well producing in-focus images but these are of course two-dimensional. How do we retrieve the 3D information from our stack of

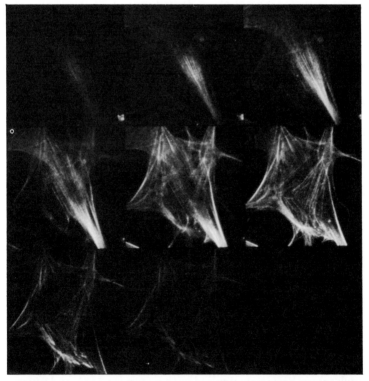

FIGURE 5.5: Montage of optical sections. Computer generated montage of confocal optical sections selected from a stack of 20 taken through a MRC5 human lung epithelial cell labeled with an actin-binding toxin coupled to rhodamine. Δz was 1.0 μm and the pixel size was 0.18 μm.

optical sections? There are two main approaches. The first is simply to display the images one after another in a sequence or side-by-side as a montage (e.g. *Figure 5.5*) so that the brain can try and make up a 3D picture from the individual sections. The second and much more powerful approach is to assist the brain by recombining the sections and displaying the reconstruction in a way that mimics one of the ways we visualize our 3D surroundings, that is, by movement.

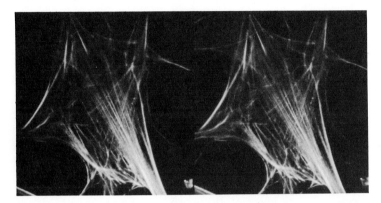

FIGURE 5.6: *Stereo projections of optical sections. Two projections through the whole stack of sections used in Figure 5.5. The projections are displayed with 6° difference between them. If viewed through a stereo viewer or by fusing the two images by eye, a 3D image should result.*

The way this is done is similar to that for reconstructing serial sections (Section 5.1). The images are stored digitally and the computer calculates what it would look like if all the sections were placed on top of one another and the stack was viewed from above. This calculation is repeated for a number of different angles, producing a set of projections [5]. Now if these projections are displayed one after another in rapid succession, say from $-20°$ through $0°$ to $+20°$, a 3D image is produced which appears to rotate around an axis (usually the y-axis). This gives the impression of viewing the specimen that has been optically sectioned from each side, greatly increasing its apparent depth. For publication, two projections can be printed side by side as in *Figure 5.6* and viewed through a stereo viewer or, with practice, by fusing the two images by eye (this process is explained in Section 10.7).

References

1. White, J.G., Amos, W.B. and Fordham, M. (1987) *J. Cell Biol.*, **105**, 41.

2. Shotton, D.M. (1989) *J. Cell Sci.*, **94**, 175.

3. Brakenhoff, G.J., van Spronsen, E.A., van der Voort, H.T.M. and Nanninga, N. (1989) *Meth. Cell Biol.,* **30,** 379.

4. Boyde, A. (1985) *Science,* **230,** 1270.

5. Agard, D.A. (1984) *Ann. Rev. Biophys. Bioeng.,* **13,** 191.

6. Agard, D.A., Hiraoka, Y., Shaw, P.J. and Sedat, J.W. (1989) *Meth. Cell Biol.,* **30,** 353.

7. Shaw, P.J. and Rawlins, D.J. (1991) *Prog. Biophys. Mol. Biol.,* **56,** 187.

6 How to Get the Best Image

There is a saying that computer people use: 'Garbage in = garbage out!'. This applies just as well to microscopy. The most sophisticated microscope equipment cannot make a poor specimen into a good one. When preparing the specimen remember to bear in mind the type of microscopy to be used and the structures to be examined; this will save all sorts of problems later at the microscope.

6.1 Preparation of unfixed specimens

One of the attractions of biological microscopy is that a tissue can often be observed directly mounted in a drop of water or buffer without any other pretreatment. The *Spirogyra* filaments shown in *Figures 8.4* and *10.2*, for example, were simply placed in a drop of DAPI solution (4',6-diamidino-2-phenylindole; a stain for DNA) and covered with a cover slip. The main consideration with an unfixed specimen is keeping it stationary long enough to observe and photograph it and, with a very thick specimen, being able to focus on the desired portion of it. To reduce movement of the specimen, if using a cover slip with a liquid mounting medium (see Section 6.2.7), it is a good idea to have just enough liquid to prevent the specimen drying out. Another approach is to use a more viscous mounting medium (e.g. solutions of glycerol, polyvinyl alcohol (PVA), polyethylene glycol (PEG) or Tissue-Tek: a mixture of PVA and PEG sold as a support medium for cryo-sectioning) although these may be toxic to the specimen. With larger specimens, gluing them to the slide or support (e.g. with cyanoacrylate adhesive) may be possible before mounting in water or buffer.

To help with focusing on a very thick specimen, ideally a long working distance objective should be used (see Section 3.1) or otherwise a normal objective with as low a magnification as is practicable (as this will also have quite a long working distance). The problem with normal objectives at medium or high magnifications is that the front lens of the objective will hit the specimen or cover slip before the desired focal plane is reached.

6.2 Preparation of fixed specimens

In most cases, some form of fixation will be required to enable the specimen to be preserved, sectioned and then stained or labeled before it can be examined under the microscope. Detailed descriptions of these procedures can be found in textbooks of histological methods [1–3], but the following sections give brief outlines of the different steps. Many of these are common to both light microscopy (LM) and electron microscopy (EM) and useful information on LM sample preparation can often be obtained from textbooks on EM sample preparation; see, for example, references [4–6].

6.2.1 Fixation

Fixation is used to denature the tissue and to preserve it in as near a lifelike state as possible. There are two main ways to fix biological material: physical methods including freezing and heating, and chemical methods.

Freezing or cryo-fixation [7–9] has the advantage that under ideal conditions, very fast moving cellular events can be stopped and recorded at an ultrastructural level (e.g. by freeze-fracture EM). However, for LM applications, for example, immunolabeling of cryo-sections, such structural and temporal resolution is not usually required and the main advantage of cryo-fixation is that antigenicity is preserved better than with chemically fixed specimens. Specimens should be small enough to provide fairly rapid cooling throughout the depth of the tissue and so as an approximate guide, pieces of tissue should not be more than 0.5 cm along any edge. Common methods of freezing are by plunging into liquid nitrogen or by spraying with an aerosol of freon (see Appendix D for suppliers). The tissue can be mounted in a support medium such as Tissue-Tek (see Section 6.1) to allow easier handling of the frozen block, or the specimen can be stuck directly onto the cryo-microtome chuck. To help prevent ice crystals forming during freezing (which would disrupt the cells), cryo-protectants (e.g. PVA or sucrose) can be infiltrated into the tissue.

Heating also fixes biological material and nowadays is often accomplished using either a domestic or a purpose-built microwave oven. This method produces rapid and uniform fixation of, for example, hospital biopsy samples, with structural preservation comparable to chemical fixation with formalin (see below). For further details, see references [10, 11].

In general, chemical fixatives cross-link and/or denature proteins, preventing their movement or function, (for more details on the mechanism of action of fixatives see references [1] and [6]). Chemical fixatives include aldehydes (e.g. formaldehyde and glutaraldehyde), alcohols (e.g. methanol or ethanol, often together with acetic acid) and oxidizing agents (e.g. osmium tetroxide). The most widely used fixative is formaldehyde often at 4% (w/v) in a suitable buffer at neutral pH. This can be diluted from a stock solution (known as 'formalin' – a 35% (w/v) solution in water) or better, made freshly from solid *para*formaldehyde. To prepare a 4% solution of formaldehyde: 8% (w/v) *para*formaldehyde is made in water and heated to 60°C in a fume cupboard. Then, one or two drops of NaOH or KOH (about 5 M) are added which will dissolve the *para*formaldehyde. After cooling, readjust the volume if significant evaporation has occurred. Finally, an equal volume of double strength buffer at pH 6.8–7.0 is added to dilute the formaldehyde to 4% (w/v). This method minimizes the amount of alkali needed to dissolve the *para*formaldehyde and so avoids having to readjust the pH of the fixative. The buffer used depends on the specific application but buffers containing chloride ions (e.g. Tris-HCl or phosphate buffered saline) should be avoided as there is the possibility that HCl vapor could be released which reacts with formaldehyde to give the very carcinogenic gas *bis*-chloromethylether. Suitable buffers for animal tissue are Sorensen's phosphate buffer (made by mixing 0.4 M $Na_2HPO_4 \cdot 12H_2O$ and 0.4 M $NaH_2PO_4 \cdot 2H_2O$ to give the desired final pH – equal volumes give pH 6.8) or for plant material, 50 mM 1,4-piperazinediethanesulfonic acid (PIPES; the acid, not the disodium salt), 5 mM EGTA and 5 mM $MgSO_4$ adjusted to pH 6.9 with KOH ('microtubule stabilizing buffer'; reference [12]).

For some tissue and cell preparations, fixation can be followed immediately by mounting the tissue for microscopy. Also, one type of microtome, a vibrating knife microtome often called by the trade name 'Vibratome', can be used on fresh or fixed tissue without embedding. For many tissues, however, embedding in a solid medium is necessary prior to sectioning.

6.2.2 Tissue handling

Tissue pieces can be placed in molds of various shapes and sizes for embedding. The molds should be resistant to the solvents used and to the heat in the polymerization process. A very wide range of sizes and shapes of mold is available either from specialist microscope accessory suppliers or from general scientific suppliers (see Appendix D).

Cell suspensions are often fixed and embedded by pelleting into a suitable tube or mold by centrifugation. An alternative for spread cell preparations is to use a special cyto-centrifuge ('Cytospin') which has special adapters to hold glass slides in such a way that cells are forced to spread out over the surface of the slide in a thin layer.

6.2.3 Dehydration and embedding

Embedding media fall into two categories: waxes and resins (plastics). Both are suitable for conventional staining and enzyme histocytochemistry, whilst immunocytochemistry is most often performed on resin sections. Waxes include paraffin, ester and polyester waxes, of which the

most commonly used is paraffin wax. The two main classes of resin are the acrylic resins and the epoxy resins. For light microscopy, acrylic resins such as LR White are commonly used as the sections are hydrophilic, so allowing staining and labeling with aqueous reagents. Antigenicity is also preserved better than with the epoxy resins. Resins specifically designed for LM are also available and include JB4, GMA, Histocryl and Nanoplast (a melamine resin); JB4 and GMA are soluble in water before polymerization. These resins can be obtained from specialist microscope accessory suppliers (see Appendix D).

Apart from the water soluble resins, embedding media are not miscible with water so the first step is to remove all the water from a tissue and replace it with something that is miscible with the embedding medium. An alcohol series is most commonly used, starting at 30 or 50% (v/v) in water and then increasing the concentration up to 100% (v/v) alcohol. With paraffin wax but not with the resins, this is followed by a 'clearing' step which substitutes a wax-miscible solvent (e.g. xylene or citrus fruit oils) for the ethanol. Dehydration protocols will vary according to the tissue but a typical protocol using ethanol might be as follows: 50% (v/v), 20 min; 70% (v/v), 20 min; 90% (v/v), 30 min; 100% (v/v) 30 min; 100% (v/v), 30 min. The specimen is placed in a suitable container so that, when embedded, the 'block' can be mounted in the microtome directly. The embedding medium is introduced gradually until the specimen is completely infiltrated. It is then polymerized by heat, UV light or in the case of paraffin wax by allowing to cool. Automatic embedding machines are available for both wax and resin embedding and are useful for the routine embedding of large numbers of specimens.

6.2.4 Sectioning

A variety of microtomes are available which will cut suitable sections for light microscopy. 'Rotary', 'rocker' or 'sledge' microtomes are designed for wax blocks and will cut sections between about 5 and 30 μm using steel knives. For resin blocks, ultramicrotomes that are designed for cutting ultrathin sections for EM are used at the top end of their range to cut 0.5–2 μm sections for LM using glass knives. Microtomes for frozen tissue (often called by the trade name 'Cryostat') are similar to those for wax-embedded tissue but are mounted inside a freezer. They will cut sections of about the same thickness (5–30 μm) using steel knives. A vibrating blade microtome ('Vibratome') uses a steel knife blade which vibrates parallel to the cutting edge while it is advanced over the tissue block. The tissue is stuck to a base plate and covered in water or buffer.

Embedded sections can be picked up on treated (see Section 6.2.5) or untreated glass slides. Paraffin wax sections can be picked up from the knife in a ribbon using forceps. The sections are then floated on a bath of warm water to spread them flat and picked up from underneath. Resin sections are floated directly onto the surface of water in a bath which is held adjacent to the glass knife edge. They can be picked up with a fine bristle (e.g. an eyelash) mounted on the end of a cocktail stick or with a wire loop (about 3 mm diameter) and placed on a drop of water on a slide. In both cases, the slides are then warmed on a hot plate at about 45°C to spread the sections flat and stick them to the slide. Cryo-sections can be picked up directly from the cooled knife blade by placing a slide at room temperature on top of them. The sections immediately melt on to the slide and can then be air dried.

6.2.5 Sticking tissue to slides

One of the most common problems with handling sections is that they float off the slide during staining or washing. There are a variety of proprietary tissue bonding agents available (new ones are coming out all the time) in addition to treatments with agents such as poly-L-lysine which have been widely used for many years; treating clean (ethanol washed) slides with a 1% (w/v) solution of poly-L-lysine is quite effective for sticking single cells and tissue sections.

One of the best of the modern bonding agents is γ-aminopropyl triethoxysilane (APTES; Sigma) which sticks the tissue down very firmly. It does not seem to give a high background with antibodies or in *in situ* hybridization procedures. A protocol for APTES treatment (modified from the original [13]) is as follows: ethanol washed slides are placed in a freshly prepared 2% (v/v) solution of APTES in acetone for 10 sec (not longer otherwise non-specific binding of e.g. antibodies becomes a problem). They are then transferred briefly to acetone alone and then air dried. This primes the slides with free amino groups ready for activation. Slides so treated can be stored for up to 6 months but gradually lose their effectiveness. Just before they are needed, the slides are placed in a 2.5% (v/v) solution of glutaraldehyde in phosphate buffer (e.g. Sorensen's phosphate buffer, pH 7.4; see Section 6.2.1) for at least 30 min. They are then rinsed in water and air dried.

6.2.6 Staining and labeling

This topic covers a very wide range of procedures outside the scope of this book and so references are given here to texts containing detailed protocols. General histochemical and immunocytochemical methods can be found in references [1–3] and [14] for animal tissue and [15 and 16] for plant tissue. Animal enzyme histochemistry is covered comprehensively in reference [17]. General chromosome staining is dealt with in reference [18] and chromosome banding in reference [19]. Fluorescence labeling of cells and tissues is covered in references [20–23] while techniques for autoradiography are described in reference [24] and for *in situ* hybridization in references [25 and 26].

6.2.7 Mounting media

In most cases, you will need to cover your specimen with a cover slip (see Section 6.2.8). Exceptions are when using a stereo microscope, very low power (up to ×4) on a compound microscope or objectives that are designed to be immersed in water, glycerol or oil without a cover slip. To support the cover slip and surround the specimen in a medium of the correct refractive index, solutions called mountants are used. The correct refractive index is important as several microscopical techniques rely on the refractive index of the medium being either very similar or very different to that of the specimen.

Mountants fall into two categories: semi-permanent (i.e. they do not set hard) and permanent (they do). The main advantage of semi-permanent mountants is that the cover slip can be washed off at a later date for re-labeling. Also, most anti-fade mountants for fluorescence microscopy (see Section 7.3.3) are semi-permanent. An obvious semi-permanent mountant is water or buffer, which are excellent for unstained specimens using phase contrast and Nomarski. These evaporate quite quickly, however, although this can be reduced by sealing the edges with rubber cement or nail varnish. Glycerol, either neat or 90% (v/v) in buffer, is also commonly used, especially for fluorescently labeled specimens. This stays moist indefinitely if kept at 4°C, especially if the edges are sealed. It has a higher refractive index than water and is less good for unstained specimens. It also tends to fade colored stains.

There are two popular permanent mountants available at the moment: Euparal and DePeX (or DPX). Canada balsam was widely used at one time and is still available but it is strongly autofluorescent. All three are dissolved in solvents which are irritants and/or harmful (DPX and Canada balsam are dissolved in xylene, which is also explosive and highly flammable). These are non-aqueous so the specimen needs to be dry before adding the mountant. Euparal has a slight yellow color which some workers prefer for stained specimens, while DPX is colorless. The properties of different mountants are summarized in Tables 6.1 and 6.2.

TABLE 6.1: *Properties of mounting media*

Mountant	Good for	Disadvantages
Semi-permanent		
Water/buffer	Living cells, unstained tissue phase contrast and Nomarski	Dries out
Glycerol or glycerol/ buffer	Medium term storage, fluorescent specimens (with anti-fade)	Fades colored stains
Permanent		
Euparal, DePeX, Canada balsam	Long term storage, wax or resin sections	May not be compatible with fluorescence

TABLE 6.2: *Refractive indices of common mounting media*

Mounting medium	Refractive index
Water	1.33
Glycerol	1.47
Euparal	1.481
New Entellan	1.493
Entellan	1.500
Immersion oil*	1.515
DePeX	1.529
Canada balsam	1.528

*Zeiss.

Whatever mountant is used, care should be taken to blot most of it out before it dries or before the edges are sealed. This is because objectives are designed assuming the specimen is right next to the underside of the cover slip. If there is a thick layer of mounting medium in between the specimen and the cover slip, it may not be possible to focus on the specimen or the image will be degraded.

6.2.8 Cover slips

Most objectives are designed to be used with a cover slip which is 0.17 mm thick. This corresponds to a manufacturer's number of 1½. The higher the NA (and usually the higher the magnification), the more critical, in theory, the tolerance around this thickness becomes. Thus at NA = 0.75 (say ×40), the tolerance is only ±4 µm before degradation of the image will occur (especially with a dry lens).

In practice, a 1½ cover slip is fine for most applications, especially as the thickness of the mounting medium may vary. You will notice a difference with a much thinner or thicker cover slip, however: the image will be unclear and lacking in contrast. The approximate thicknesses of the different cover slip numbers are shown in Table 6.3. With oil immersion

TABLE 6.3: *Cover slip thicknesses*

Manufacturer's number	Thickness (mm)
0	0.085–0.13
1	0.13–0.16
1½	0.16–0.19
2	0.19–0.25
3	0.25–0.35

objectives, the cover slip thickness is not so important except at very high NA. However, 1½ coverslips can be recommended generally unless there is a special need for a different thickness.

6.2.9 Immersion media

The immersion medium is the substance between the objective and the cover slip or, if a cover slip is not present, the specimen itself. In the latter case, the semi-permanent mountants described in Section 6.2.7 (i.e. water and glycerol) can also be used as immersion media provided that the objective has a correction collar for immersion media of different refractive indices (see Section 3.1). In most cases, however, a cover slip is used and in this case, one of the proprietary immersion oils is ideal. Most have a refractive index of approximately 1.515 although for specialist applications, including critical reduction of spherical aberration [27], a range of immersion oils with different refractive indices can be obtained (see Appendix D). Immersion oils from different manufacturers should not be mixed as most combinations will degrade the image and some will autofluoresce.

6.3 Use of microscope controls

6.3.1 Use of objectives

• Objectives can cost up to £2000 so be very careful with them. If you remove one of the objectives, never leave it on the bench but always replace it in its protective case to prevent it getting knocked, the front lens scratched or dust getting into the back.
• Make sure the lens surface is clean. Especially make sure that when using a dry lens, it is clean of any immersion oil. See Section 6.4 for the best method to clean a lens; *do not* use any type of alcohol or acetone as this can dissolve the cement holding the lenses in. Also check there is no dust on the back lens (where this lens is visible).
• Use the correct immersion oil, generally refractive index (η) = 1.515. Do not mix immersion oils from different manufacturers. If there is a correction collar, check that it is set to the right position. To apply immersion oil, focus down the objective until there is a gap of about 2 cm between the front lens and the cover slip. With a dropper bottle or pipet add a small drop of oil, taking care not to introduce air bubbles. Enough is needed so that when the slide is brought up to meet the objective there is a continuous film of oil over the whole surface of the front lens. If there is insufficient oil, part of the image will be in focus and a large bubble will appear over the rest which will move around if the focus is adjusted. However, do not put so much oil on that it drips over the edge of the slide onto the stage mechanism, condenser, etc. Apart from being messy, immersion

oil is corrosive to some plastics and the paint on some microscope parts.

• Some objectives have a locked up position. When these are used, make sure they are in the down position for two reasons. First, they are designed to work that way and second, if by chance you focus too close and hit the cover slip, the lens will compress against a spring instead of driving itself through the slide.

• Lenses from a single manufacturer will often be parfocal, i.e. there is no need to change the focus controls when switching between different objectives. However, unless you know the objectives well, do not assume this as you may end up crunching the lens into the slide or sweeping off the cover slip when you turn it round. Always lower the stage first before changing objectives, and then refocus.

• Lastly, remember to put the slide the right way up. This sounds rather obvious but it is very easy to put the slide upside down and wonder why the microscope will not focus on the specimen.

6.3.2 Use of the condenser

• Keep the top lens clean.

• For the best results at high magnification, immersion oil can be put between the condenser and the underneath of the slide (oiling the condenser). As it is rather messy, it is not a very common practice but while it is probably not beneficial for lower NA objectives, it really does give a substantial improvement at NA ≥ 1 (see Section 7.1.4).

• Should it be necessary to remove the condenser, for example, to clean immersion oil off the top lens, lower the condenser as far as possible before undoing the clamp. Be careful not to scratch the top lens as it is removed. It may be necessary to raise the stage to produce enough clearance and this may involve taking off some of the objectives.

• If the condenser centering controls are adjusted so that the condenser is a long way off the optical axis of the microscope, no light will reach the specimen and so the microscope will be very difficult to set up for any type of illumination. If this is suspected, remove the condenser and examine the position of the condenser body in its holder from underneath. Adjust the centering controls until the body of the condenser is approximately central in the holder, and then replace on the microscope.

6.3.3 Use of eyepieces

• Once again, keep clean; dirt is very noticeable on the eyepieces so, next to the objectives, they are the most important to keep clean. One or both may have a correction collar to compensate for different people's eyesight. To start off with, set both to the same position (usually 0) and then focus the specimen (see Section 6.3.5).

• Remember not to squint but to relax your eyes (try to look into the distance as if you were looking at a far away building). If you have problems seeing down both eyepieces at once, look at *Figure 6.1*. You have to adjust

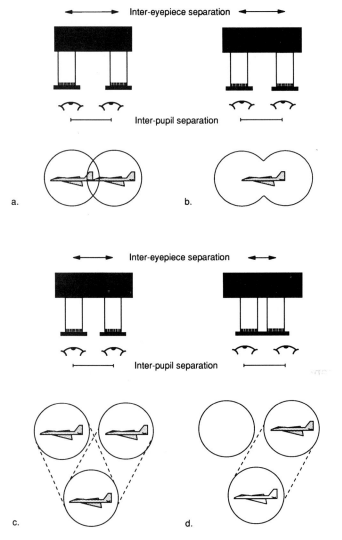

FIGURE 6.1: Eyepiece separation. The correct separation is shown in **c**.

the separation between the two eyepieces so that the light coming out of each forms an image on the retina of each of your eyes that your brain can fuse into one (*Figure 6.1c*). If they are too far apart (*Figure 6.1b*), you will get an image like looking through binoculars on a 1950s black and white film! Even further apart (*Figure 6.1a*) and you will see two images which will be almost impossible to fuse. If the eyepieces are too close together (*Figure 6.1d*), you will see only one of the images while your other eye will be distracted by objects in your peripheral vision. If you have tried to adjust the distance and failed, as a last resort measure the distance between your eyes with a ruler then set this distance between the binoculars.

- If you wear glasses you may have problems seeing into the eyepieces properly. The first remedy is to remove your glasses and adjust the correction collar (see below). If this is not satisfactory, you may require so-called high eye-point lenses, which produce the image further away from the top of the eyepiece than normal. Eyepiece hoods to shield stray light may also be useful.

• If you are using a monocular microscope, an eye patch is useful to save having to keep the other eye closed or else putting up with distractions in your peripheral vision.

• At higher magnifications, reflections from the lenses in the eyepiece may mean that you will see images of your eyelashes or even the capillaries in your retina. This can be very distracting until you get used to it.

• Once you have a binocular image more or less in focus for both eyes, you can adjust the eyepiece correction collars to the correct settings for each eye individually. If there is an eyepiece graticule in one of the eyepieces, for example a photoscreen for the camera, close your other eye and adjust the eyepiece correction collar until the graticule is sharply in focus. Be careful not to squint too hard or you will distort your eye and mess up the focus. Check the specimen is still in focus and make an adjustment with the specimen fine focus if necessary. If the specimen and the graticule are in focus with this eye, change eyes and adjust the other eyepiece until the specimen is in focus. Do a final check with both eyes separately. You should now have everything in focus in both eyes. It is essential to check your eyepieces are properly focused to produce in-focus photographs (see Section 10.2).

• You may want to insert a graticule in the eyepiece for measuring or counting. This occupies the same place as the photoscreen (if fitted) which is usually on a ledge below or between the eyepiece lenses. Remove the eyepiece to see if there is a suitable ledge. If there is, unscrew the relevant piece of tube and place the graticule on the ledge. Make sure it is clean, the right way up and, in microscopes with a predetermined eyepiece orientation, at the correct angle so the scale or grid is horizontal when you look down the microscope. Suppliers of graticules are listed in Appendix D.

6.3.4 Use of the specimen stage

• Many microscopes have stages with controls for adjusting the movement of the whole stage, or a portion of it, in the x and y directions. Vernier scales are often fitted so that a particular position of the stage can be recorded, or for use in making measurements on the specimen (see also Chapter 9).

• Some microscopes allow rotation of the stage, and in this case there are adjustment screws for centering the stage around the optical axis of the microscope. Stage rotation is useful for composing a good photograph and is essential for polarized light microscopy (see Section 7.6). To center the stage, undo the locking nut and, with a low magnification objective focused on a specimen, rotate the stage backwards and forwards by about ⅛ turn. The image will probably move sideways or up and down as well as rotate.

Now turn one of the adjustment screws and repeat the rotation to see if the x, y movement has been reduced. If it has made it worse, try moving the adjustment screw the other way. Only turn the screw ¼–½ a turn each time. Now try the other screw. By trial and error it should eventually be possible to center the stage so that it does not move, just rotates around a single point. Now increase the magnification (say ×63) and repeat using (much) smaller turns of the adjustment screws. Reckon to spend at least 15 min on this process—it cannot be hurried. The results are useful and rather pleasing!

• If you need to remove the stage for any reason, take the condenser off first to give plenty of room in which to work. There is usually just a clamp to undo and then the stage will come away from the stand.

6.3.5 Use of the focus controls

• First position the specimen under the objective. This sounds obvious but the most common reason for not being able to focus on the specimen is because it is outside the light beam. With a low contrast specimen this can be difficult. One tip is to close either the field or the condenser apertures down to give a small point of illumination on the slide. Now move the slide backwards and forwards across the light beam and look at the top of the slide (not down the eyepieces). It should be possible to see the specimen as it crosses the beam; at that point it is under the objective. Remember to open the aperture up again afterwards. Alternatively, focus on the specimen with a low magnification lens which will cover a much wider field, and then change to a high power lens.

• Next move the stage up towards the objective using the *coarse* focus control until the objective is about *2 mm* away from the cover slip or, if using an immersion lens, until the oil 'jumps' onto the objective. Continue using the coarse focus control if you can move it slowly enough (it sometimes needs a fairly strong wrist) and look for a changing image. A change in intensity or color may be seen well before you start to see an in-focus image. Now use the fine focus for the final adjustment to focus on the specimen. Focus gently at all times and be aware of any resistance or sound indicating the cover slip is in contact with the front of the objective.

• The reason for using the coarse focus as much as possible is not only because it is quicker. Quite a number of microscopes have a limited travel on the fine focus so that if you only use this control, it will eventually come up against a stop. On some microscopes, the focus will stop operating a few turns before this happens, at which point there will be no further change in the image focus. If either of these occur, *do not continue to turn the fine focus control in that direction*: serious damage can result. Instead, turn it the other way several turns or until the focus begins to move again and then focus back with the *coarse* focus control. Repeat this a few times to centralize the fine focus control on its travel.

6.3.6 Bulb centering

Only tungsten or tungsten/halogen bulbs will be considered here – consult your microscope handbook for mercury bulb centering. It is often surprising to find how much the image is improved after centering the bulb. It is important with all forms of microscopy to get an even illumination across the whole field of view. This cannot be done if the bulb is badly out of alignment. Some of the more modern microscopes have pre-centered bulbs with no provision for adjustment. If you have a handbook for your microscope, consult it for instructions on how to center the bulb. If not, the following is suitable for most microscopes: place a piece of thin white paper (e.g. lens tissue) over the field aperture and turn on the lamp. Most lamps have a ground glass diffusing screen – if you can, move this out of the way so that an image of the bulb filament is projected onto the paper. Adjust the lamp focus control until the image is sharp. If this is not possible, simply look at the brightest part of the image. Now adjust the centering controls until the image is as central and as bright as possible and replace the diffuser.

If you are using an external lamp, the bulb will probably be centered with respect to the field aperture (if fitted) and so no adjustment will be needed apart from setting up Köhler illumination (see Section 7.1) If there is not an aperture, follow the instructions above, adjusting the position of the lamp and the mirror to get the brightest and most central image of the bulb possible.

6.3.7 Use of the Bertrand lens or phase telescope

Use of these devices to adjust the phase annulus for phase contrast is described in Section 7.2.2. They can usually be focused and so can also be used to inspect intermediate image planes. For example, *Figure 6.2* shows oil on the front lens of a dry objective and *Figure 6.3* shows dirt on the back

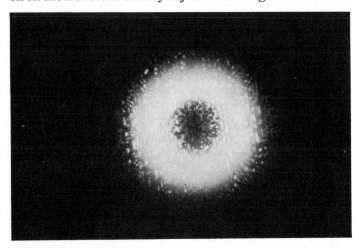

FIGURE 6.2: *Use of the Bertrand lens. An oil smear on a dry objective examined using the Bertrand lens.*

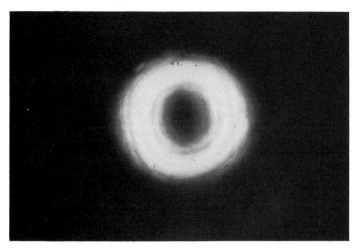

FIGURE 6.3: *Dust on the back lens of an objective.*

lens of an objective. First insert the Bertrand lens into the light path by pushing in or rotating its holder. Then focus it until a sharp image comes into view. This will look like a dark tube. To see which element is in focus, try moving it and observe what happens to the image. For the objective, partially unscrewing it immediately shows whether this is the element which is in focus. If focused on the condenser aperture plane you can check that the condenser aperture itself is centered. On some microscopes it can be adjusted like the phase annulus.

6.4 Maintenance

Routine maintenance involves cleaning lenses and other optical surfaces of dust, immersion oil, grease and other contaminants. These activities are vital to get good images as there are a lot of optical surfaces between the specimen and your eyes and any dirt will seriously degrade the image.

- Dust can be removed using either one of the commercially available compressed clean air canisters, a photographer's puff duster or clean lens tissue. See Appendix D for suppliers.

- Light contamination of the lens surface can be removed by 'hahing' and wiping with clean lens tissue. Always wipe in one direction only—never in a circular motion as this may cause any particles of glass dust from slides, cover slips, etc. to be ground into the lens. Use a different part of the tissue for each wipe.

- For heavier contamination, clean lens tissue and a suitable solvent should be used with the same technique as above. Although opinion varies widely about the best solvents for cleaning lenses, it is generally accepted that whatever solvent is used, it should be used sparingly as it may leak between the lens elements and possibly dissolve the cement that is used to mount the lenses in their cases. Acetone and most alcohols are definitely *not recommended* as cleaning solvents as they are known to attack some of these cements. The best solvents are water for aqueous contaminants (e.g. dried on stain) or analytical grade diethyl ether (poorer grades leave a film on the lens) for everything else. Xylene is also commonly used, but it is more toxic than diethyl ether. Do not use any solvent in an unventilated room and do not leave the top off the bottle. Be very careful when using ether as it is extremely flammable. Keep it away from all sources of ignition (including starting a mercury lamp). Ideally, and certainly if a lot of cleaning work is to be done, do everything in a fume cupboard.

● Cleaning the immersion oil off objectives (even immersion ones) is important as, if kept in contact for long periods, oil can seep between the elements of the objective which will ruin it.

● Always replace microscope dust covers after use and keep the area around the microscope clean. Do not leave objectives or eyepieces where they can be knocked over but always store them in the appropriate containers.

Most other maintenance should be left to a qualified engineer. If a microscope needs to be partially dismantled for cleaning, be careful to reassemble it in exactly the reverse order. If you have to use a screwdriver, you would probably be advised to leave it alone!

References

1. Bancroft, J.D. and Stevens, A. (1990) *Theory and Practice of Histological Techniques,* 3rd Edn. Churchill Livingstone, Edinburgh.

2. Kiernan, J.A. (1990) *Histological and Histochemical Methods: Theory and Practice,* 2nd Edn. Pergamon Press, Oxford.

3. Horobin, R.W. (1989) in *Light Microscopy in Biology: A Practical Approach* (A.J. Lacey, ed.). IRL Press, Oxford, p. 137.

4. Roland, J.C. and Vian, B. (1991) in *Electron Microscopy of Plant Cells* (J.L. Hall and C. Hawes, eds). Academic Press, London, p. 1.

5. Glauert, A.M. (1977) in *Practical Methods in Electron Microscopy, Vol. 3* (A.M. Glauert, ed.). North-Holland, Amsterdam, p. 1.

6. Hayat, M.A. (1981) *Fixation for Electron Microscopy.* Academic Press, London.

7. Roos, N. and Morgan, A.J. (1990) *Cryopreparation of Thin Biological Specimens for Electron Microscopy: Methods and Applications,* Royal Microscopical Society Handbook No. 21. Oxford University Press, Oxford.

8. Robards, A.W. (1991) in *Electron Microscopy of Plant Cells* (J.L. Hall and C. Hawes, eds). Academic Press, London, p. 257.

9. Robards, A.W. and Sleytr, U.B. (1985) in *Practical Methods in Electron Microscopy, Vol. 10* (A.M. Glauert, ed.). Elsevier, Amsterdam, p. 1.

10. Boon, M.E. and Kolk, L.P. (1989) *Microwave Cookbook of Pathology*, 2nd Edn. Coulomb Press, London.

11. Hopwood, D. (1992) *R.M.S. Proc.*, **27**(2), 71.

12. Lloyd, C.W. and Wells, B. (1985) *J. Cell Sci.*, **75**, 225.

13. Tortellote, W.M., Schmid, P., Pick, P., Verity, N., Martinez, S. and Shapshak, P. (1986) *Neurochem. Res.*, **12**, 265.

14. Clark, G. (1981) *Staining Procedures*, 4th Edn. Williams and Wilkins, Baltimore.

15. Gahan, P.B. (1984) *Plant Histochemistry and Cytochemistry: An Introduction.* Academic Press, New York.

16. Harris, N. (1992) *J. Micros.*, **166**, 3.

17. Stoward, P.J. and Pearse, A.G.E. (1991) *Histochemistry: Theoretical and Applied, Vol. 3, Enzyme Histochemistry*, 4th Edn. Churchill Livingstone, Edinburgh.

18. Darlington, C.D. and La Cour, L.F. (1969) *The Handling of Chromosomes*. George Allen & Unwin, London.

19. Sumner, A.T. (1989) in *Light Microscopy in Biology: A Practical Approach* (A.J. Lacey, ed.). IRL Press, Oxford, p. 279.

20. Wang, Y.-L. and Taylor, D.L. (eds) (1989) *Methods in Cell Biology*, Vol. 30. Academic Press, London.

21. Rost, F.W.D. (1991) *Quantitative Fluorescence Microscopy*. Cambridge University Press, Cambridge.

22. Becker, E. (1985) *Fluorescence Microscopy*. Leica UK Ltd, Milton Keynes.

23. Holz, H.M. (1985) *Worthwhile Facts About Fluorescence Microscopy*. Carl Zeiss (Oberkochen) Ltd, Welwyn Garden City.

24. Baker, J.R.J. (1989) *Autoradiography: A Comprehensive Overview*, Royal Microscopical Society Handbook No. 18. Oxford University Press, Oxford.

25. Polak, J.M. and McGee, J.O'D. (eds) (1991) In situ *hybridization. Principles and Practice*. Oxford University Press, Oxford.

26. Leitch, A.R., Schwarzacher, T., Jackson, D. and Leitch, I.J. (in press) *A Practical Guide to* in situ *Hybridization*, Royal Microscopical Society Handbook. Oxford University Press, Oxford.

27. Hiraoka, Y., Sedat, J.W. and Agard, D.A. (1990) *Biophys. J.*, **57**, 325.

7 Setting Up the Microscope

This chapter describes in practical terms the operation of the different imaging modes described in Part 1. For further background information, see references [1–4].

7.1 Bright field (Köhler) illumination

7.1.1 Equipment required

This technique requires a microscope equipped with an objective (preferably several of different magnifications, for example, ×10, ×20, ×40 and ×63 or ×100), a condenser (a simple 'Abbe' type is fine for bright field) and if photographs are needed, a camera setup. Most newer microscopes come with built-in light sources. However, if an external light source is used, it should preferably have an adjustable aperture placed close to the bulb or collector lens. This will be used as the field aperture for setting up the illumination.

For high magnification work the condenser should have a numerical aperture (NA) matching that of the highest NA objective to be used. For low magnification, modern condensers are designed so that the top lens swings out when the condenser is lowered, allowing a wider field of illumination. Others have an accessory lens which is swung in when using lower power.

7.1.2 Setup procedure

Bright field forms the basis of all other types of transmitted light microscopy except dark field, so it is essential that you can do it quickly and accurately. Setting up bright field illumination is described below and in the flow diagram in *Figure 7.1* and photographs of the various steps are shown in *Figure 7.2a–f*. The lens used was a Zeiss ×40/0.9 Ph3 oil immersion lens and the specimen was an acetocarmine-stained metaphase preparation of chromosomes of the broad bean *Vicia faba*. You may need to refer to *Figure 3.1* to help locate the different parts.

BRIGHT FIELD ILLUMINATION

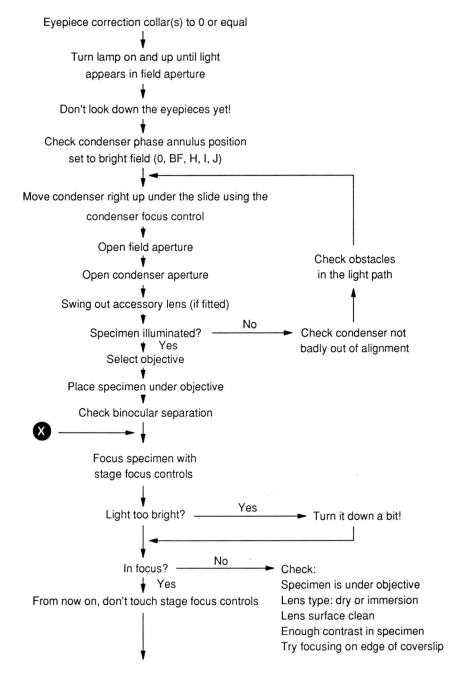

Eyepiece correction collar(s) to 0 or equal

↓

Turn lamp on and up until light
appears in field aperture

↓

Don't look down the eyepieces yet!

↓

Check condenser phase annulus position
set to bright field (0, BF, H, I, J)

↓

Move condenser right up under the slide using the
condenser focus control

↓

Open field aperture

↓

Open condenser aperture

↓

Swing out accessory lens (if fitted)

↓

Specimen illuminated? ———No———▶ Check condenser not
↓ Yes   badly out of alignment

Check obstacles
in the light path

Select objective

↓

Place specimen under objective

↓

Check binocular separation

X ———————▶ ↓

Focus specimen with
stage focus controls

↓

Light too bright? ———Yes———▶ Turn it down a bit!

↓

In focus? ———No———▶ Check:
↓ Yes   Specimen is under objective
From now on, don't touch stage focus controls   Lens type: dry or immersion
                                                 Lens surface clean
                                                 Enough contrast in specimen
                                                 Try focusing on edge of coverslip

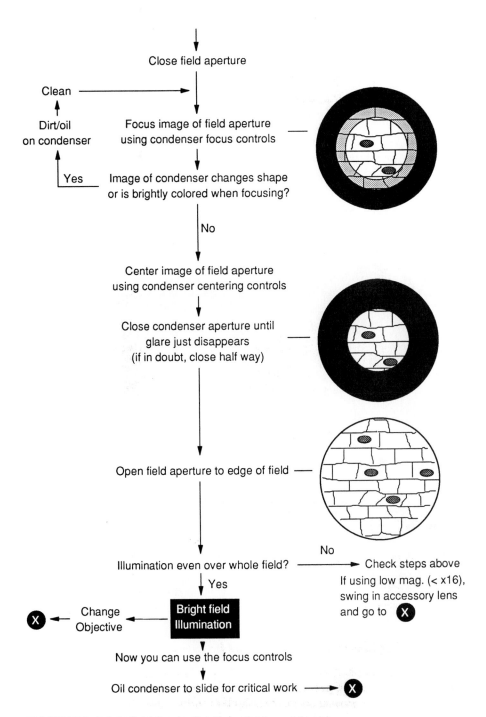

FIGURE 7.1: *Bright field illumination flow-chart.*

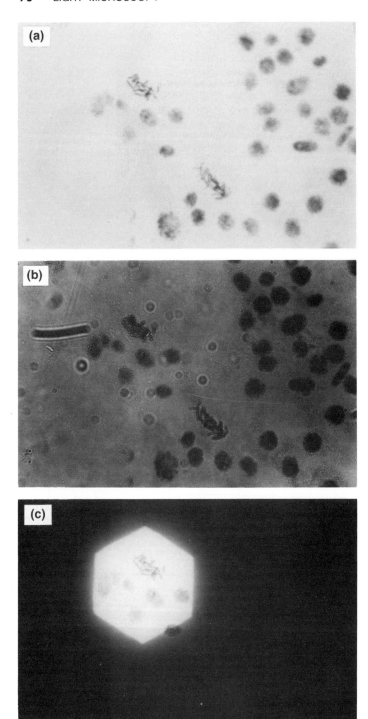

FIGURE 7.2: Setup procedure for Köhler illumination. **a.** Focus the specimen.
b. Close the field aperture. **c.** Focus the condenser until the field aperture is sharp.

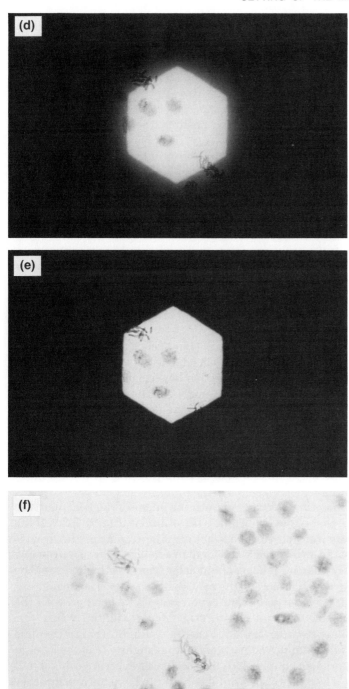

FIGURE 7.2: Continued. **d.** Center the condenser. **e.** Adjust the condenser aperture.
f. Open the field aperture to give bright field illumination.

- Set the eyepiece correction collar(s) to 0 or so that they are both on the same setting. Do not look down the eyepieces yet.
- Turn on the lamp and turn up the voltage until light can be seen in the field aperture. Put the prepared slide on the stage.
- Turn the phase annulus selector (if fitted) to the bright field position (usually indicated by 0, BF, H, J or I).
- Using the *condenser focus* control, position the condenser right up against the underneath of the slide.
- Open both the *field* and *condenser apertures* fully.
- Swing out the accessory lens and filter holder if fitted.
- Light should now be seen shining onto the slide from the condenser. If not, go back and check that all the previous steps have been done correctly. If light is still not seen, there may be some obstacle in the light path or the condenser may be way out of alignment (see Section 6.3.2).
- Once the slide is illuminated, select an objective and position the specimen underneath it. If it is an immersion lens, add some immersion oil or the appropriate medium (see Sections 6.2.9 and 6.3.1). Check the separation of the eyepieces and focus the specimen (see Section 6.3.5). If the illumination is too bright, use *only* the lamp intensity control to dim the light.
- If it is not possible to focus on the specimen, check that it is correctly positioned underneath the objective. Check the lens type again (dry or immersion). Is the surface of the objective lens clean? Is there too little contrast in the specimen? If the specimen is near to being in focus try focusing on the edge of the cover slip then moving back to the specimen.
- Let us assume there is now an in-focus image, albeit not very pretty (*Figure 7.2a*). From now on, *do not touch the stage focus controls*. Adjust only the controls affecting the *condenser*.
- First close the *field aperture*. The image will probably go dark (*Figure 7.2b*).
- Adjust the *condenser focus* control until the image of the field aperture is sharp. It will probably have a bright halo around it (*Figure 7.2c*). If the image of the condenser aperture does not remain circular as the focus is changed, but instead swings back and forth in an ellipse, there is probably dirt or oil on the condenser lens. A slight color fringe or the image of the aperture blurring which occurs as the condenser is focused is normal.
- If the image of the aperture is off center, adjust its position with the *condenser centering controls* until it is in the middle (*Figure 7.2d*).
- Adjust the *condenser aperture* until the halo of light around the field aperture image *just* disappears. Now open it up slightly (*Figure 7.2e*). If the halo cannot be seen, simply close the condenser aperture control half way. This will be a bit too far open but it is better to have too much rather than too little aperture. An alternative method for setting the condenser aperture is to examine the back focal plane of the objective using a Bertrand lens or phase telescope (see Section 6.3.7). Close the aperture until it just comes into view against the bright circle of light. Now close it a little further: to about ⅓ closed. Now return to normal viewing. Both

methods give approximately the same result but the former is slightly quicker and easier. For further details on setting the condenser aperture, see Section 7.1.5.

● Now open the *field aperture* until it just disappears outside the field of view (*Figure 7.2f*). This is now bright field *or* Köhler illumination. Only when the condenser adjustments are completed should the stage focus controls be used.

7.1.3 Low power observation

If all the steps in Section 7.1.2 have been followed and the illumination is not uniform across the field of view but brighter in the middle than at the edges, it is either because the lamp has not been adjusted properly (see Section 6.3.6) or because a low (×16 or less) magnification lens is being used. In the latter case, if the top lens of the condenser is of the type that can be swung out of the light path, do so. Otherwise swing in the accessory lens (if fitted) and readjust the condenser. At very low magnifications (e.g. ×2.5), even this will not illuminate the whole field of view (*Figure 7.3a*). All that can be done is to lower the condenser right *down* and open up both apertures fully. This is not Köhler illumination but at least the whole specimen will be visible (*Figure 7.3b*).

Another problem that can occur with low power observation is that an image of the ground glass screen can coincide with that of the specimen and field aperture, giving a speckled pattern. In this case, try swinging in the accessory lens and readjusting the condenser. If the problem persists, defocus the condenser slightly until the speckled pattern disappears.

7.1.4 Oiling the condenser

At very high magnifications (×63 or more), it is worth putting immersion oil between the condenser and the underneath of the slide (oiling the condenser). With the specimen focused, lower the condenser until there is room to get an oiler under the slide. Place some oil on the top lens of the condenser (more will be needed than for an objective) and then bring the condenser up until the oil jumps onto the slide. Now adjust the condenser as above. A considerable improvement should be noticed in both the image quality and its brightness.

7.1.5 Effect of the condenser aperture

As explained in Section 3.2, setting the condenser aperture involves compromising between resolution and contrast. The best resolution is obtained by matching the condenser aperture exactly with the NA of the objective. This is achieved by closing the aperture until it just appears in view against the illumination (whilst looking at the back focal plane of the objective). The problem with this is that at low to medium magnification,

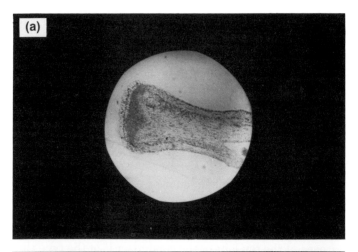

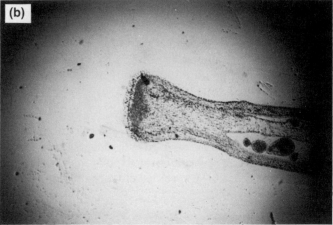

FIGURE 7.3: *Low power observation.* Brassica napus *embryo.* **a.** *The field of view is severely restricted under Köhler illumination.* **b.** *Focusing the condenser right down and opening both apertures allow the specimen to be seen.*

this results in very little contrast in the image. As contrast is often more important than the best possible resolution, a condenser aperture setting is suggested which is rather more closed than this. Other texts vary on the setting of the condenser aperture between ¼ and ½ closed. The first method described in Section 7.1.2 produces a setting about ½ closed which gives good contrast at low and medium magnifications. At high magnification and high NA (especially > 1.3), to get the best resolution, an equivalent high NA condenser is required and the condenser aperture should be set to match the NA of the objective as described above.

The effect of having the condenser aperture wrongly set is demonstrated in *Figure 7.4a* and *b*. *Figure 7.4c* shows the image produced with the condenser aperture set correctly. If the aperture is closed right down, an image that superficially looks like it has more contrast will be produced

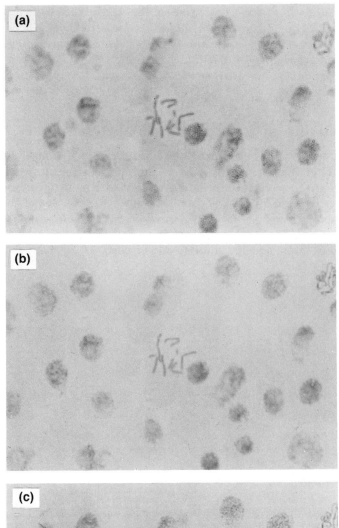

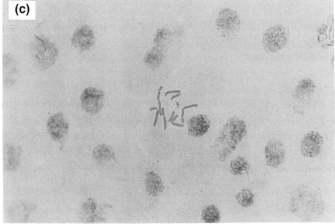

FIGURE 7.4: *Effect of condenser aperture.* **a.** *Condenser aperture closed. Note the haloes around all the features.* **b.** *Condenser aperture fully open reducing the contrast.* **c.** *Correctly set up bright field illumination.*

(*Figure 7.4a*). However, upon closer inspection this can be seen to be due to haloes around the image and in fact the resolution is badly degraded. On the other hand, opening the aperture right up will give a bright image which lacks contrast (*Figure 7.4b*). Although attempts should always be made to set the condenser aperture correctly, it is better to have it more rather than less open.

7.1.6 Colored filters

If the specimen is only faintly stained, the contrast can be increased by using a colored filter. This is inserted into the light path either by swinging it in underneath the condenser or by selecting it with a button on the microscope base. Use a color that is complementary to the color of the stain. For example, *Figure 7.5* shows the same specimen as *Figure 7.4c* (which is stained magenta with acetocarmine) but with a green filter selected. For a blue stain, a yellow filter should be used.

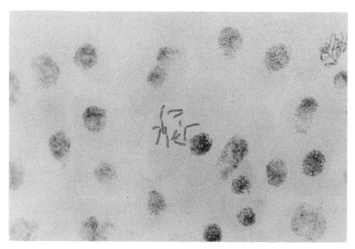

FIGURE 7.5: *Use of colored filters. The addition of a green filter improves the contrast slightly compared with Figure 7.4c.*

It was seen in Section 2.2 that the wavelength (color) of light affects the resolution in the image. In theory, the smaller the wavelength, the better the resolution. However, chromatic aberration of the lenses in the objective (see Section 3.1) also has a significant effect and, in practice, using apochromatic lenses, green light gives the best resolution.

7.1.7 Troubleshooting

This section highlights some of the common faults that occur with bright field microscopy. *Figures 7.6–7.8* are of a resin section through an embryo of *Brassica napus* (oil-seed rape) stained with toluidine blue.

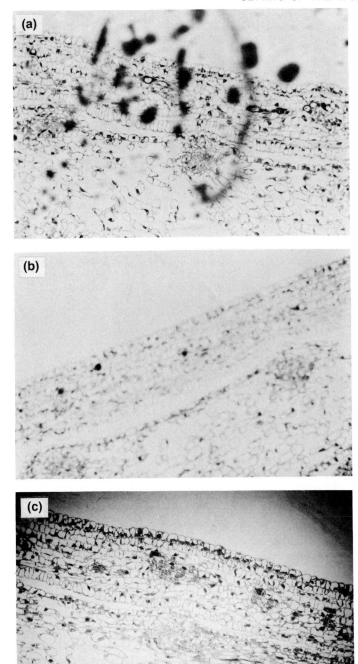

FIGURE 7.6: *Troubleshooting: using a dry lens.* **a.** *Dirt on the cover slip.* **b.** *Slide upside down. The specimen cannot be focused.* **c.** *The condenser is incorrectly focused and there is no accessory lens, giving very uneven illumination.*

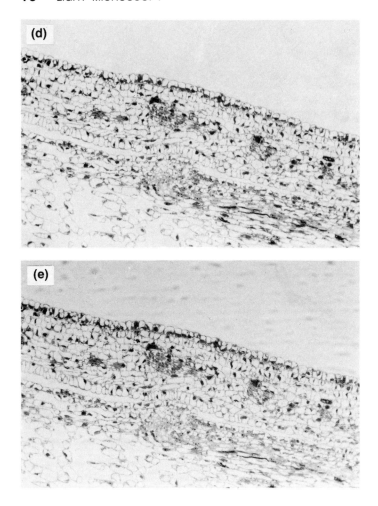

FIGURE 7.6: *Continued.* **d.** *A speckled pattern can be formed if the condenser is not properly focused.* **e.** *Again, the condenser is incorrectly focused and immersion oil can be seen underneath the slide.*

Figures 7.6 and *7.7* illustrate some problems with dry lenses ($\times 16$ in this case). Compare the pictures to one obtained with a correctly set up microscope (*Figure 7.7d*). In particular note the following: in *Figure 7.6c* and *d* the source of the speckled pattern is the ground glass screen in the lamp housing, visible because the condenser is not properly focused; in *Figure 7.6e* the condenser is still incorrectly focused and a smear of immersion oil is visible on the underneath of the slide. The obstruction in *Figure 7.7a* is a fluorescence supplementary filter holder in the field of view. In both *Figures 7.6b* and *7.7c* the slide is upside down and, although in the latter the specimen is in focus, contrast is reduced compared to the correctly set up specimen and microscope.

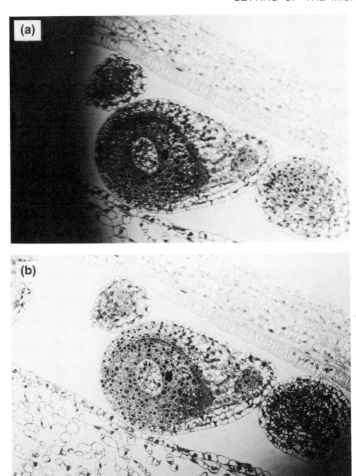

FIGURE 7.7: *Troubleshooting: more problems using a dry lens.* **a.** *There is an obstruction in the light path (a filter holder).* **b.** *The condenser is not centered, resulting in very uneven illumination.*

Figure 7.8 illustrates some of the problems that can arise with immersion oil (or lack of it). In *Figure 7.8a* there is immersion oil on a dry lens. The remaining figures in this series were obtained using a ×25 oil immersion lens. Compare *Figure 7.8e,* in which immersion oil has been correctly applied, with *Figure 7.8c* in which there is no oil either on the lens or the slide, and with *Figure 7.8b* in which there is oil on the lens but not enough to form a continuous layer between the lens and the cover slip. Note the enhanced contrast only when everything is as it should be (*Figure 7.8e*). Finally, *Figure 7.8d* illustrates one of the big problems with immersion oil – bubbles!

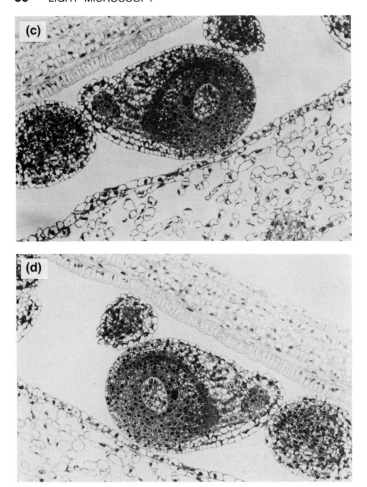

FIGURE 7.7: *Continued.* **c.** *The slide is upside down. Although in focus, there is less contrast than in* **d**, *which is correctly set up.*

7.2 Phase contrast microscopy

7.2.1 Equipment required

The extras required other than a microscope as described in Section 7.1 are objectives containing phase plates (phase objectives) and a phase condenser with matching rings to those in the objectives. If these are all bought from the same manufacturer they will all match, but those from different manufacturers may not. For example, a Ph3 objective from one manufacturer may match with a Ph2 annulus in the condenser from another manufacturer, or not at all.

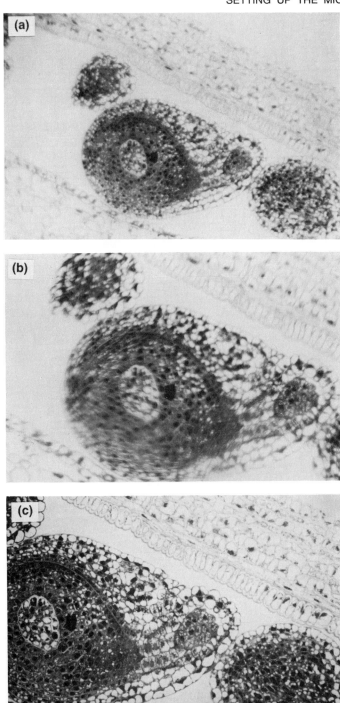

FIGURE 7.8: *Troubleshooting: problems with immersion oil.* **a.** *There is immersion oil on the (dry) lens.* **b.** *An immersion lens with oil on the lens but none on the slide.* **c.** *An immersion lens but no oil anywhere.*

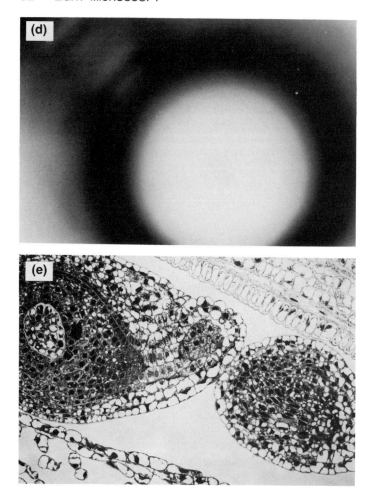

FIGURE 7.8: *Continued.* **d.** *Bubbles.* **e.** *The correct use of oil.*

7.2.2 Setup procedure

Phase contrast uses bright field illumination with a phase annulus and phase plate in the light path. These need to be exactly aligned in order to work properly. Immersion media with low refractive indices are good for phase contrast as they accentuate the difference between the specimen and the background. The steps needed to set up phase contrast illumination are described below and in *Figure 7.9*. Photomicrographs of the different stages are shown in *Figure 7.10a–f*, again using a ×40/0.79 Ph3 oil immersion lens and acetocarmine-stained chromosomes of *Vicia faba*.

- Set up the microscope for bright field illumination (see Section 7.1.2) using a phase objective. These have the letters 'Ph' and a number e.g. Ph3 written on them. The number refers to the size of the phase ring inside the objective. Ph3 is normally found on higher magnification lenses and Ph2 on intermediate and low magnification ones. If there is insufficient contrast in the specimen to see it under bright field, one of two things can be done. First, set up the illumination as described below with a stained slide then replace it with your unstained specimen. Secondly, focus on the edge of the cover slip, move away a little and set up the illumination as below.

Once the phase contrast is set up, the specimen can then be located. *Figure 7.10a* shows the specimen properly illuminated under bright field.

- Select the appropriate phase annulus by rotating the phase annulus selector on the condenser. The number should correspond to that of the objective. Thus using a Ph2 objective, select phase annulus 2. There may be 'Ph2', 'Ph40', etc. marked at one of the positions.
- Close the field aperture and refocus the condenser (*Figure 7.10b*). More colors around the image of the aperture will be seen than with bright field; this is normal. An important point to note is that if there is a condenser aperture present when a phase annulus has been selected, it *must* be fully open or nothing will be seen. There may be a mark on the aperture control 'Ph' which corresponds to the fully open position. Alternatively, the aperture control may disappear when the phase annulus is selected, thus overcoming the problem.
- If the phase annulus is adjustable, open the field aperture and inspect the back focal plane of the objective (where the phase plate is located) using the Bertrand lens or with a phase telescope. The Bertrand lens will probably have 'Ph' on it and there may be a slider to push in. If there is neither of these, use a phase telescope which is like an eyepiece but which has an extra sliding barrel. To use this, take out one of the eyepieces and replace it with the telescope. With either device a blurred ring of light will probably be seen at this stage (*Figure 7.10c*).
- Focus the Bertrand lens or the telescope so that the phase annulus and phase plate ring can be seen clearly (*Figure 7.10d*).
- Using the phase annulus adjustment controls on the condenser, adjust the position of the bright ring (the condenser phase annulus) until it is superimposed on the dark ring (the objective phase ring; *Figure 7.10e*). The reason why the objective phase ring appears dark, although it retards light less than the surrounding plate, is because it is deliberately darkened during manufacture to compensate for the different overall intensities produced by the two areas. Be careful not to touch the centering controls for the condenser itself. The phase annulus adjustment controls may involve pushing two rods into the condenser and turning them to move the annulus. Another system uses hexagonal nuts recessed in two holes on either side of the condenser or at the back. A third system has a knurled wheel at the front of the condenser and a slider at the side for adjusting the annuli.

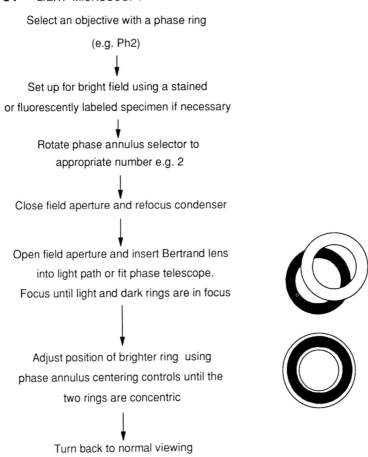

Select an objective with a phase ring

(e.g. Ph2)

↓

Set up for bright field using a stained

or fluorescently labeled specimen if necessary

↓

Rotate phase annulus selector to

appropriate number e.g. 2

↓

Close field aperture and refocus condenser

↓

Open field aperture and insert Bertrand lens

into light path or fit phase telescope.

Focus until light and dark rings are in focus

↓

Adjust position of brighter ring using

phase annulus centering controls until the

two rings are concentric

↓

Turn back to normal viewing

↓

Phase contrast

FIGURE 7.9: *Phase contrast flow-chart.*

• Now turn back to normal viewing or replace the eyepiece. A phase contrast image of the specimen should now be seen (*Figure 7.10f*).
• Contrast can be increased further by using a green filter to color the illumination.

7.2.3 Advantages and disadvantages of phase contrast

Comparing *Figures 7.10a* and *f*, the advantages of phase contrast over the equivalent bright field image are immediately apparent. *Figure 7.11a* is an unstained 1 μm resin section through an embryo sac of *Brassica napus*. The outlines of the cells are clear and the nuclei appear dark. Compare this with the equivalent bright field image (*Figure 7.11b*) and the advantages are obvious. As a second example of phase contrast on unstained

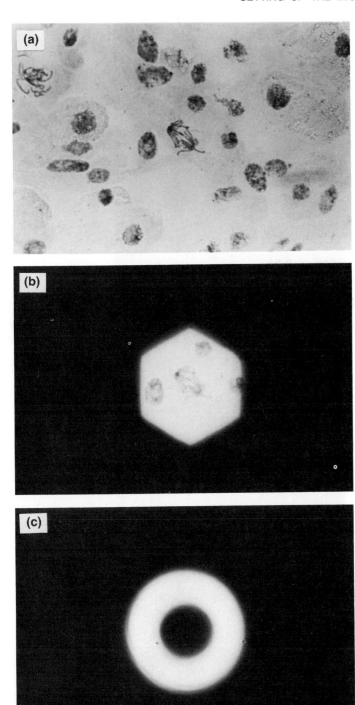

FIGURE 7.10: *Phase contrast setup.* **a.** *Bright field image.* **b.** *Field aperture closed.* **c.** *Bertrand lens moved into the light path.*

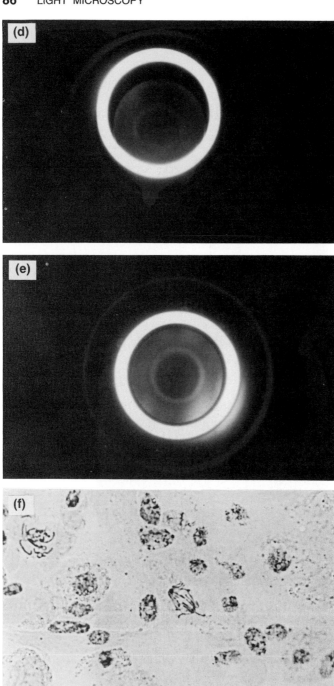

FIGURE 7.10: Continued. **d.** Phase rings in-focus and out of alignment. **e.** Phase rings aligned. **f.** Phase contrast.

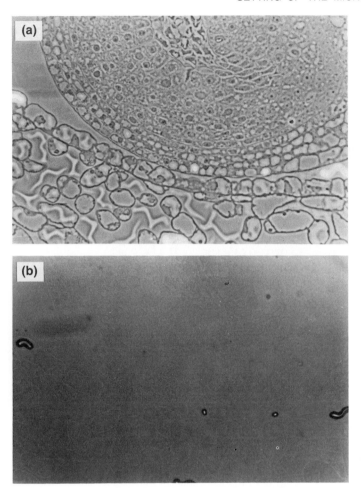

FIGURE 7.11: *Comparison of phase contrast and bright field imaging.* **a.** *Unstained resin section of a* Brassica napus *embryo sac under phase contrast.* **b.** *The corresponding bright field image.*

specimens, *Figure 7.12* is a high magnification picture of human metaphase chromosomes. Here, a ×100/1.3 Ph3 oil immersion lens was used with an oiled condenser and the chromosome preparation was mounted in Euparal. The chromosomes were completely invisible using bright field alone.

Phase contrast images are often colored but this should not be interpreted as being real color in the specimen. It is an artifact, although quite a useful one. Less wanted artifacts of phase contrast are bright haloes around dark structures. These are formed as light is redistributed throughout the image and mean that the resolution obtained is less than in bright field. It also means that phase contrast microscopy is unsuitable for thick specimens as these haloes multiply with thickness, giving a confused image.

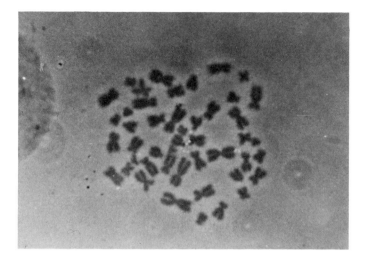

FIGURE 7.12: *Phase contrast image of unstained human chromosomes.*

7.3 Fluorescence microscopy

7.3.1 Equipment required

A microscope equipped with a mercury lamp epi-illumination system is needed together with appropriate fluorescence filter sets for the fluorochromes you want to use. These will normally include either UV, blue or green excitation filters with the appropriate dichroic mirrors and emission filters for, for example, 4′,6-diamidino-2-phenylindole (DAPI), fluorescein isothiocyanate (FITC) and tetramethyl B rhodamine isothiocyanate (TRITC). Details of the filter sets for these fluorochromes can be found in Appendix C. A phase contrast system is also very useful. The ideal is to be able to do all transmitted light modes and fluorescence on the same microscope.

7.3.2 Choice of objective

If using UV illumination, UV transmitting objectives will be required as many other types of lenses (but not all) do not transmit UV light. With epi-illumination, the brightness of the image seen is proportional to NA to the fourth power, therefore it is obviously a good idea to use as high numerical aperture objectives as possible. These will usually be immersion lenses. However, with very brightly labeled tissue, or even sometimes with autofluorescence, a good result can be obtained even with a ×2.5/0.08 NA objective.

7.3.3 Anti-fade mountants

Choice of mounting media is important with fluorescence microscopy as the light given off by fluorochromes fades quite rapidly. To reduce this fading or 'bleaching', anti-fade mountants are usually used. Bleaching is thought to be caused by oxygen radicals which allow the conversion of the fluorochrome from one energy state to another which does not give off light. Therefore, most anti-fades are oxygen radical scavengers. A popular proprietary one is 'Citifluor' which is especially good for FITC. This contains diazabicyclo[2.2.2]octane (DABCO), which is corrosive and so should be used with the appropriate precautions. Most of the other anti-fade agents are also harmful and some workers prefer to seal the cover slips with nail varnish or rubber cement before use.

An alternative, nontoxic anti-fade that is especially good for rhodamine is propyl gallate (5% (w/v) tumbled overnight to dissolve in 90% (v/v) glycerol/Tris buffered saline). Both this and Citifluor are good for DAPI. Specimens mounted in anti-fade will retain their ability to fluoresce for many months; however, the anti-fade properties of the mountant do not last as long. Therefore it is worth soaking off the cover slip (being careful not to get residual immersion oil from the cover slip onto the specimen; see below) and remounting (this cannot be done if the edges have been sealed).

7.3.4 Practical hints

- Bulb alignment is especially important with fluorescence microscopy and it is advisable to consult the manufacturer's instructions. Some types of mercury bulb get appreciably dimmer with age. It is a good idea to keep a logbook to record bulb usage (see Section 3.8): in this way the operator can change the bulb promptly after it has passed its useful life.
- Focusing can be a problem with fluorescence microscopy as there is usually a dark background with (perhaps) some small bright spots. With experience, it is possible to recognize how close the specimen is to being in focus by slight changes in the background illumination. Alternatively, a low power objective or phase contrast can be used to locate the specimen initially. Focusing on the edge of the cover slip is another possibility.
- With semi-permanent slides, for example, those mounted in one of the glycerol-based anti-fade mountants mentioned above, never get mounting medium mixed with immersion oil. If this does occur, blot it off straight away as the two do not mix optically. It is best to seal the edges of the cover slip with nail varnish to avoid this possibility. It is invariably a disaster if immersion oil actually comes into contact with a specimen in an aqueous medium. Sometimes this can be remedied by soaking off the cover slip in buffer then rinsing with *neat* Tween 20 then copious amounts of water. This drastic procedure does not seem to affect the fluorescence significantly but it is definitely a last resort technique.
- Some fixatives (e.g. glutaraldehyde) make tissues become autofluorescent. One way round this is to treat the fixed tissue with freshly prepared

solutions of 0.1% (w/v) sodium borohydride (caution – toxic and highly flammable) in phosphate buffered saline (12 mM NaCl, 16 mM Na_2HPO_4, 8 mM NaH_2PO_4, pH 8.0). Four treatments of 15 min each works well for plant tissue.

• Stains for bright field microscopy are generally incompatible with fluorescence as they tend to quench the emitted light. So, for instance, do not counterstain with toluidine blue if DAPI is going to be used.

• Finally, one should be wary of bleed-through from one fluorochrome to another. Bleed-through (*Figure 7.13*) can happen when observing specimens double labeled with, say, DAPI for nuclear DNA and FITC-conjugated antibodies. If the DAPI fluorescence is examined first, blue light will be emitted and so the nuclei will appear blue. Unfortunately, some of this blue light will also excite the FITC molecules attached to any antibodies lying near to the nucleus. This will start to cause bleaching of the FITC molecules (see Section 7.3.3) while the blue light from the DAPI is being observed. On changing to the FITC channel, a bleached area may be seen where the nuclear DAPI staining was. To avoid this problem when working with double-labeled specimens, try to work from the longest excitation wavelength to the shortest, that is, in the above example, look at the FITC fluorescence first then the DAPI fluorescence. Do not look at the DAPI fluorescence for too long if you want to go back to FITC again.

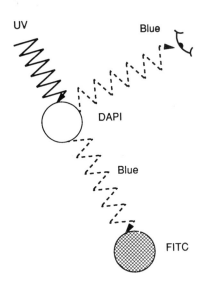

FIGURE 7.13: *Fluorescence bleed-through. Blue light emitted from DAPI molecules fades adjacent FITC molecules.*

• As the spectra of the commonly used fluorochromes are quite close to one another (see *Figure 4.6*), the filters used to separate out, for example, the emissions of fluorescein and rhodamine cannot completely cut out the light from the shorter wavelength fluorochrome. This means that sometimes an FITC image will be seen as well as a rhodamine image. If double labeling is attempted, it is better to use fluorochromes with greater spectral separation (e.g. FITC and Texas Red).

7.4 Dark field illumination

7.4.1 Equipment required

In addition to a bright field microscope, a dark field condenser is required. This can simply be a home-made patch stop below the ordinary condenser (see below) or a purpose made one fitted into the same ring that holds the phase annuli. There are also the purpose made dark field condensers as described in Section 4.4.

7.4.2 Setup procedure

There are two important points to remember. Firstly, oblique illumination is necessary. If the NA of the objective is too large in comparison with that of the illuminating rays, some of the incident non-reflected light will get into the objective, such that a dark background will not be produced and instead a distorted bright field image will result. To help produce the oblique illumination, it is best if the condenser is oiled to the slide (see Section 7.1.4) except for one of the reflection types of condenser (*Figure 4.10b*). Secondly, because only oblique illumination is used, the condenser must be centered exactly on the optical axis of the microscope.

- Start with a bright field condenser or no condenser at all and focus on the specimen.
- If the condenser has a dark field stop built in, select it. If using the appropriate type of condenser, apply oil between the condenser lens and the slide and bring the condenser up towards the slide until the oil jumps. It is very important to *open the condenser aperture* otherwise nothing will be seen. Turn the light up to maximum and focus the condenser until the best dark field is obtained. This is when the background is black, the reflective objects are bright and the illumination is as uniform as possible.
- If a purpose built dark field condenser is used, it will have to be centered. This is a more difficult operation. Start as before with an in-focus bright field image then remove the ordinary condenser and attach the dark field condenser. Turn the light right up and look into the eyepieces. Select a Bertrand lens (if fitted) or fit a phase telescope to allow the back focal plane of the objective to be observed. Focus the Bertrand lens and then focus the *condenser* up and down until a circular patch appears that is darker than the rest. At this point, adjust the condenser centering controls until this dark patch is central and then return to normal viewing or replace the eyepiece. Little will be discernible at this stage. Oil the condenser, if appropriate, as above. Now, looking down the eyepieces, adjust the condenser until the dark field is achieved. Finally, the condenser may need fine adjustment to ensure that it is central – concentrate on the position of the darkest part of the image as the condenser is focused, and adjust until it is central.

7.4.3 Practical hints

• With high NA objectives it may not be possible to achieve a dark background unless the NA can be reduced – some objectives have an iris in them for this purpose. Adjust this and the condenser focus until the best dark field image is produced.

• At low magnification, the highest number phase annulus (3 or 4) may work as a patch stop if the condenser does not already have one. It will probably be necessary to oil the condenser to the slide.

• A patch stop can be improvised very effectively by putting a disk of cardboard underneath a bright field condenser (there is often a filter holder under the condenser which is ideal), supported on a piece of clear plastic (e.g. an overhead transparency sheet). Center this by moving the disk around. The correct size of the stop can be determined by trial and error – too big and there will not be enough illumination, too small and a dark background will not be produced. Remember to oil the condenser to the slide. This apparently crude method can be surprisingly effective.

• The oil immersion type of condenser is the most difficult to center but gives the best results, especially at high magnification. However, the cup shaped reflective type is fine at low magnification.

• If you use a counterstained specimen, e.g. toluidine blue, the small amount of incident light getting into the objective may be enough to show this up. This can be useful to outline cell boundaries, etc.

7.5 Nomarski (DIC) microscopy

7.5.1 Equipment required

Because of the complex nature of the elements for Nomarski microscopy, a purpose designed Nomarski system is essential. A rotating stage is also useful for orienting the specimen with respect to the 'direction' of illumination. One of the beam splitters is normally put in the condenser where it can be selected like a phase annulus. The other is often put directly above the objective (and matched to it) but is sometimes higher up and may be combined with the analyzer.

7.5.2. Setup procedure

• Properly set up Köhler illumination is critical for Nomarski and the condenser aperture has a much greater effect on the contrast than with bright field. Refer to *Figure 4.16* for a diagram of the different components. First insert the two polars (polarizer and analyzer) and cross them to give a dark background.

• Select the appropriate beam splitter in the condenser for the objective being used and adjust the lateral position of the second beam splitter by rotating a nut until the contrast is satisfactory. Most systems produce a dark band at 45° to the field of view which is movable by adjusting the second beam splitter. The contrast can also be varied with the condenser aperture although the same rules apply as for bright field (Sections 7.1.2 and 7.1.5).

- Plan–apo lenses should be used if at all possible.
- The orientation of the specimen is important as the same structure can look very different if turned around relative to the apparent 'direction' of illumination. However, by rotating the stage, particular features can be oriented so that they are best contrasted against the background.
- The best position of the dark band depends on the specimen but it should be adjusted so the background is as light as possible while still giving relief.
- Birefringent specimens interfere with Nomarski optics and are therefore not recommended with this type of microscopy.
- The optical sectioning properties of Nomarski illumination will be obvious immediately. However, these images cannot easily be reconstructed in the same way as confocal optical sections.

7.5.3 Interpreting the image

Nomarski is very good for detecting abrupt changes of refractive index that occur over a small distance. Thus, for example, nuclear membranes, lipid droplets and mitochondria are very clearly picked out. However, it is important to realize that *the apparent three-dimensional structure is misleading*. To explain this point, in *Figure 7.14* the Nomarski image of the two specimens shown would appear identical, as the interaction with the pairs of beamlets is the same over small distances. Another confusing aspect of Nomarski optics is whether a vesicle, for example, appears as a depression or as raised relief. This depends on the refractive index of the contents: if it is less than the surrounding medium, it will appear as a

a.   b.

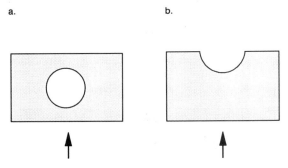

FIGURE 7.14: *Problems with interpreting Nomarski images. **a.** Shows a spherical droplet inside a medium of higher refractive index. **b.** Shows a hollow depression in the surface of the same medium. Both specimens would appear identical using Nomarski illumination.*

depression, while if it is greater it will appear raised up. Always focus up and down repeatedly to find the relationship of one structure to another. Be aware, however, that as this is done, the relative refractive indices between the object and its medium may change and so alter the appearance! Finally, the appearance of the structures underneath the dark band is not the same as dark field. More details on Nomarski image interpretation are given in reference [5].

7.6 Polarized light microscopy

7.6.1 Equipment required

A microscope with a rotating stage is required with a slot for an analyzer above the specimen and a position for the polarizer under the condenser (see *Figure 4.14*).

7.6.2 Setup procedure

- Set up for normal bright field microscopy with a birefringent specimen.
- Now add the polars and turn one (usually the polarizer) until the background goes dark (the specimen may or may not go dark as well).
- Now rotate the stage. Regular variations in intensity should be seen every 45° with a biological specimen. *Figures 7.15a* and *b* show a cell wall fragment of a stem pith cell of *Datura stramonium*. *Figure 7.15a* is at an orientation 45° to the orientation of the cross d polars. The birefringence in the specimen makes it appear light.
- If a compensator (a device that introduces a known retardation at a measured angle) is available you can try ad isting this to 'extinguish' (make fully dark) the specimen and then reac off the angle. This can be used to quantify the amount of retardation by the specimen along each of the two axes of refractive index. *Figure 7.15b* shows the same specimen as before but with a compensator introduced into the light path. As this is rotated, the specimen darkens and the background lightens. By measuring the amount and direction of rotation needed to 'extinguish' the section, the direction of birefringence (positive or negative) and the amount of retardation can be calculated (see Section 4.5).

7.7 Reflected light illumination

7.7.1 Equipment required

For epi-polarization microscopy an epi-illumination system (e.g. a fluorescence microscope), two polars, a heat filter and a 50:50 mirror are

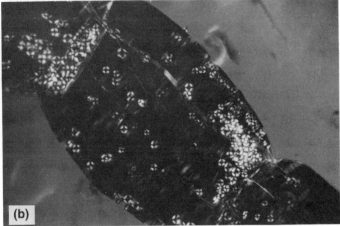

FIGURE 7.15: *A birefringent specimen (cell wall fragment of* Datura stamonium) *illuminated with polarized light.* **a.** *The specimen is at 45° to the orientation of the crossed polars so appears bright against a dark background. Note the characteristic appearance of starch grains.* **b.** *Using a compensator, the light from the specimen is extinguished (turned dark).*

required. For reflection contrast microscopy, a special objective fitted with a quarter wave plate (antiflex objective) and a patch stop condenser are also needed.

7.7.2 Setup procedure

Instructions for epi-polarization only are presented; for reflection contrast, see the manufacturer's instructions.

• Replace one of the fluorescence dichroic mirrors with a 50:50 mirror (be very careful not to scratch the thin metal layer coating the mirrors). An excitation filter should be used with a mercury bulb, to cut out harmful

UV light. However, a filter on the emission side is not required as the reflected light is the same color as the incident light. The color of the incident light is not particularly important but blue light seems to be better than green for silver particles in autoradiographic samples.

• Place one polar in the incident light path (between the lamp and the 50:50 mirror) with a heat filter on the lamp side to protect it and the other one in the reflected light path (between the 50:50 mirror and the eyepieces).

• With the specimen on the stage and preferably with an oil immersion lens, focus the specimen first with bright field or phase contrast illumination as described in Sections 7.1.2 and 7.2.2. Then turn off the transmitted light and turn on the epi-illumination. A bright background of the same color as the incident light will probably be seen, with some poorly contrasted, brighter spots, again of the same color. Now rotate one of the polars until the background goes much darker while the reflected light from the silver grains, etc. stays at about the same intensity. There will be one position (when the two polars are 'crossed') which will give the darkest background.

7.8 Confocal and three-dimensional microscopy

7.8.1 Equipment required

Most confocal microscope systems come ready to attach to a standard fluorescence microscope and the manufacturers will supply the necessary adapters. Some points to consider when buying such equipment are:

• The choice of laser will affect which fluorescent dyes can be used. The cheapest argon–ion lasers have two emission lines at 488 and 514 nm. 488 nm is fine for fluorescein but 514 nm is rather too short for good excitation of rhodamine. There are also problems with double labeling (e.g. with fluorescein and rhodamine) because as the emission lines are quite close, the fluorescence filter sets cannot be very specific without cutting out some of the incident light. As the rhodamine signal is weak compared to the fluorescein (because the excitation is not optimal) and as the fluorescence spectra of the two dyes overlap to some extent, the fluorescein image may also be seen when looking at the rhodamine one. A better laser for confocal microscopy is a krypton laser with lines at 488, 568 and 647 nm. This gives much better rhodamine excitation and also allows much more specific filters to be used, so improving double labeling. The long wavelength line is good for exciting fluorochromes like phycoerythrin. UV laser confocal microscopes are available from some of the confocal microscope manufacturers but are very expensive because of the cost of the lasers themselves and the special optics necessary to handle the UV light. For further discussion of the choice of lasers see reference [8].

● A variable confocal aperture is essential as confocality must invariably be sacrificed for image intensity in some situations. By opening up the confocal aperture slightly, a very much brighter image will be produced with only a small loss of confocality.

● Accurate control of the stage movement via the fine focus control is also essential. Some microscopes have better fine focuses than others in this respect. A stepping motor is usually used to control the fine focus although other methods (for example piezo drives) are available. The stepping mechanism and fine focus control should ideally be capable of reproducibly controlling the stage in steps 0.1–0.2 µm.

● All confocal microscope systems are now PC or work-station based. The controlling software should be easy to use and versatile enough for the operator's needs. If image analysis/processing capability is required, compare the confocal microscope manufacturer's products with those from specialist image handling software companies (see Appendix D). It may be that it is better to collect the images on the PC and then transfer them to another system for analysis and/or processing. Increasingly, companies are getting together to market jointly products to do these jobs.

7.8.2 Practical hints

● Specimen preparation is the same as for standard fluorescence microscopy (see Section 7.3). If double labeling with two fluorochromes is to be performed, apart from the laser considerations above, remember also to look at the longer wavelength fluorochrome first to prevent bleedthrough.

● An anti-fade mountant should be used to prevent bleaching of the fluorochrome. However, bleaching is no more of a problem here than with nonconfocal imaging.

● The higher the magnification, the smaller the area that will be scanned and so bleaching in this region will increase.

7.8.3 Optical sectioning

● It is important to consider how many optical sections to take and at what interval in the z-axis (Δz). Ideally, Δz should be as small as possible so as to derive the maximum amount of information from the specimen. In practice, however, the number of optical sections is likely to be limited by the amount of storage space available on the computer and its power for subsequent processing. A 512×512, 8 bit image takes up about 0.25 Mbytes of memory or disk space. Taking sections say every 0.2 µm of a 20 µm thick labeled specimen would produce 100 sections or 25 Mbytes of data. This is *big*. Unless access to formidable computing power is available, 20 sections at a coarser section interval (say 1 µm) is more reasonable (5 Mbytes). However, this does risk ignoring some of the information in the specimen and this should be borne in mind when interpreting the data. For more information on the choice of section spacing see references [7, 8].

• The choice of magnification is discussed in Sections 10.4 for film and 11.4 for video and the same considerations apply here: increasing the magnification more than that necessary to give a pixel size less than 0.2 μm will not produce greater resolution. Also, because of the amount of computer storage space taken up by images, it is not a good idea to make the image larger than necessary.

• For further discussion of the practical aspects of confocal microscopy see references [7, 8].

References

1. Lacey, A.J. (1989) in *Light Microscopy in Biology: A Practical Approach* (A.J. Lacey, ed.). IRL Press, Oxford.

2. Bradbury, S. (1989) *An Introduction to the Optical Microscope,* Royal Microscopical Society Handbook No. 1. Oxford University Press, Oxford.

3. Spencer, M. (1982) *Fundamentals of Light Microscopy.* Cambridge University Press, Cambridge.

4. Taylor, D.L. and Wang, Y.-L. (1989) *Fluorescence Microscopy of Living Cells in Culture, Vols A and B.* Academic Press, San Diego.

5. Padawer J. (1968) *J. R. Micros. Soc.,* **88,** 305.

6. Fricker, M.D. and White, N.S. (1992) *J. Micros.,* **166,** 29.

7. Pawley J.B. (ed.) (1990) *Handbook of Biological Confocal Microscopy.* Plenum Press, New York.

8. Shaw, P.J. and Rawlins, D.J. (1991) *Prog. Biophys. Mol. Biol.,* **56,** 187.

8 Case Studies

In this chapter are four examples of how viewing a single specimen with several different microscopical techniques gives enhanced information over using just a single technique. The examples also serve as illustrations of the sorts of images that can be obtained with each of the individual imaging modes.

8.1 Mitotic chromosomes: labeling the ends

Figure 8.1 shows mitotic chromosomes of the broad bean *Vicia faba* labeled by *in situ* hybridization with a digoxygenin-labeled DNA probe to telomeric sequences (the DNA at the ends of chromosomes [1]). The probe was detected using an antibody to digoxygenin and this amplified using a secondary antibody conjugated with fluorescein. A Leitz ×63/1.4 Plan–apo oil immersion objective was used for the fluorescence images and a Zeiss ×63/1.3 Ph3 oil immersion objective for the phase contrast with the condenser oiled to the slide. The phase contrast image is shown in *Figure 8.1a* and illustrates the use of this technique for visualizing specimens with very little inherent contrast. A DAPI image of the total DNA is shown in *Figure 8.1b* and the fluorescein-labeled probe in *Figure 8.1c*. By comparison of the DAPI and FITC images, the precise position of the telomeric sequences on the ends of the chromosome arms can be determined.

8.2 Floral apex: detection of mRNA

Figure 8.2 shows a wax section through a floral apex of a wild-type flower of *Antirrhinum majus* labeled by *in situ* hybridization to the mRNA of the gene *floricaula* involved in floral development [2]. The probe was a digoxygenin-labeled RNA probe derived from *floricaula* and it was detected by an alkaline phosphatase-conjugated anti-digoxygenin antibody. When the substrates (5-bromo-4-chloro-3-indolyl phosphate and nitro blue tetrazolium) of the alkaline phosphatase enzyme are added to

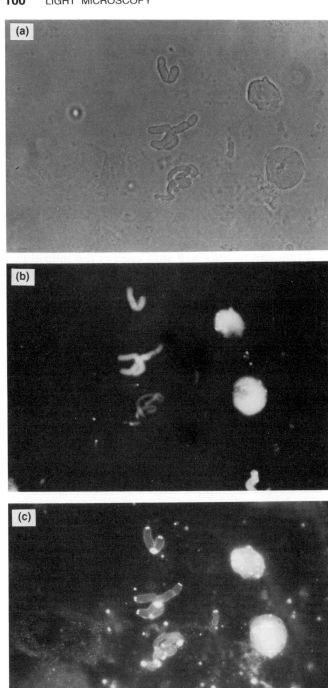

FIGURE 8.1: Vicia faba *chromosomes labeled with a probe to telomeres.* ***a.*** *Phase contrast.* ***b.*** *DAPI (for DNA) and* ***c.*** *FITC for the labeled telomeres.*

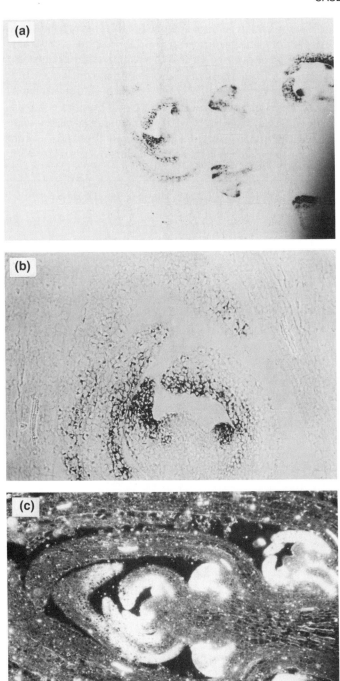

FIGURE 8.2: Floricaula *mRNA localized in a wild-type apex of* Antirrhinum majus. **a.** *Bright field.* **b.** *Phase contrast.* **c.** *Dark field.*

the specimen, a blue crystalline precipitate is produced. This can be seen in *Figure 8.2a* (a bright field image) and *Figure 8.2b* (a phase contrast image). The labeled RNA is expressed in the bract and sepal primordia and is seen as dark staining. Phase contrast also shows the dark staining and also helps to delineate the cell walls. Under dark field illumination the appearance is particularly dramatic (*Figure 8.2c*). The blue crystals are very reflective and show up strikingly with dark field. The tissue was also counterstained with a fluorescent dye (Calcifluor) that glows blue under the bright illumination. This can be seen in color in *Figure 10.3*

8.3 Human lung epithelial cells: cytoskeletal networks

Figure 8.3 shows MRC5 human embryonic lung epithelial cells which were double labeled with an antibody against microtubules and a toxin

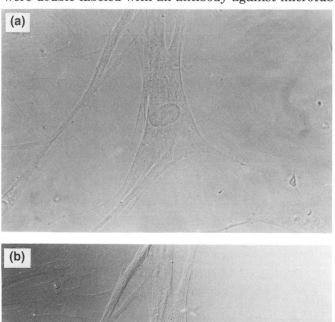

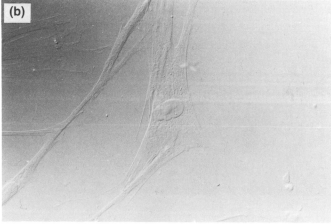

FIGURE 8.3: *MRC5 lung epithelial cell.* **a.** *Phase contrast.* **b.** *Nomarski.*

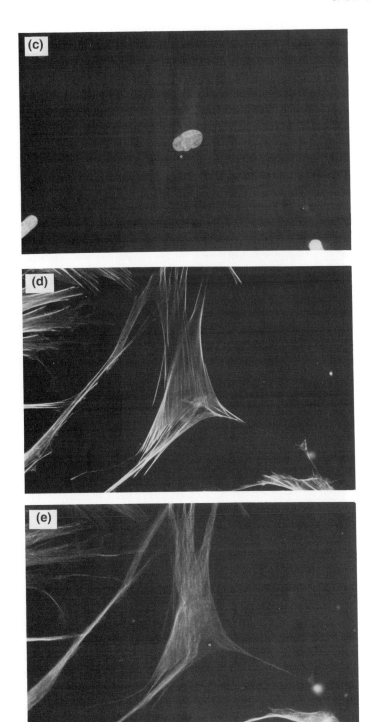

FIGURE 8.3: *Continued.* **c.** *DAPI (for DNA).* **d.** *Rhodamine–phalloidin-labeled actin-binding toxin.* **e.** *FITC labeled microtubule network.*

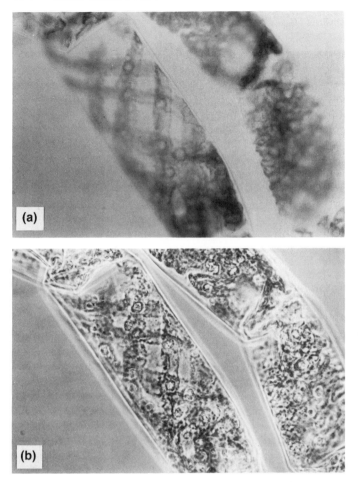

FIGURE 8.4: *Filament of* Spirogyra grevilleana. *a. Bright field. b. Phase contrast.*

that binds actin filaments. The first antibody was detected with a secondary antibody labeled with fluorescein and the actin-binding toxin directly conjugated with rhodamine. The cells were also stained with DAPI for DNA. *Figure 8.3a* shows the appearance under phase contrast. The central cell is fairly well contrasted against the background and the nucleus is clearly visible; some dark granules in the cytoplasm can also be seen. Using Nomarski illumination (*Figure 8.3b*), contrast is increased due to the shadowing effect. The processes at the ends of the cell, the nucleus and cytoplasmic granules are all easily visualized. Also, there is a hint of some fibers inside the cell running parallel to the main axis. The DAPI image (*Figure 8.3c*) shows the position of the nucleus very clearly and some substructure is discernible inside it. The rest of the cell is only very faintly visible. The rhodamine image of the actin component of the cell (*Figure 8.3d*) shows that the fibers seen with Nomarski are probably actin – the so-called stress fibers. Finally, *Figure 8.3e* demonstrates the

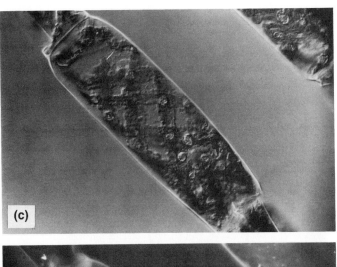

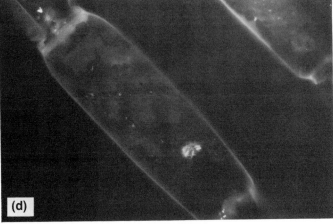

*FIGURE 8.4: Continued. **c.** Nomarski. **d.** DAPI.*

microtubule network which is distinct from the stress fibers and comes together at a focus over the nucleus. Another cell labeled in this way is shown in color in *Figure 10.1*. The lenses used for these pictures were Zeiss ×40/1.0 Plan–apo oil immersion and ×40/0.9 Ph3 Neofluar oil immersion objectives.

8.4 Pond alga: a model light microscope specimen

Figure 8.4 is of a cell of the filamentous alga *Spirogyra grevilleana* [3] grown in culture. This filamentous alga has very characteristic spiral chloroplasts. *Figure 8.4a–c* are useful to compare the differences between bright field, phase contrast and Nomarski on a fairly thick specimen. The specimen was prepared by simply staining a filament with DAPI and then mounting it in anti-fade mountant. The bright field image (*Figure 8.4a*) has little contrast, with only the outline of the chloroplasts visible. Under phase contrast illumination (*Figure 8.4b*), the chloroplasts are highly contrasted but there is a noticeable halo around the structures, particularly obvious at the cell wall. Nomarski, on the other hand, has as much, or even more, contrast so that tiny organelles can be clearly seen without a halo (*Figure 8.4c*). For transmitted light microscopy, therefore, Nomarski is the method of choice for this specimen. Using epi-fluorescence, the nucleus and some of the mitochondria can be seen (*Figure 8.4d*).

In summary, each method gives different but complementary information and it is good practice to use at least two different methods of examining any specimen.

References

1. Rawlins, D.J., Highett, M.I. and Shaw, P.J. (1991) *Chromosoma,* **100,** 424.

2. Coen, E.S., Romero, J.M., Doyle, S., Elliot, R., Murphy, G. and Carpenter, R. (1990) *Cell,* **63,** 1311.

3. Jordan, E.G. and Rawlins, D.J. (1990) *J. Cell Sci.,* **95,** 343.

9 Measuring Down the Microscope

Light microscopy is used for two types of measurement, counting and determination of lengths, areas, etc. While there is now a wide variety of computer-based systems using video cameras available, many workers still rely on non-automated methods. This chapter briefly describes the more commonly used of these methods; for more details see references [1–5].

9.1 Counting

For counting, for example, blood cells, fungal spores, pollen grains, or cells in culture, a hemocytometer is used [1]. This was originally designed for counting blood cells and is a slide marked off in a central grid area separated from the rest of the slide by two channels. It has a heavy cover slip which must be held firmly against the slide so that Newton's rings can be observed on either side of the grid area. In this configuration a counting chamber of known and constant volume is formed above the grid. The cell suspension is introduced into the chamber by capillary action from a pipet placed against the edge of the cover slip (see *Figure 9.1*). Phase contrast illumination is useful for colorless cells. By counting the number of cells (often with a tally-counter) within the grid, cell density (number per ml) can be calculated. An example of such a calculation is shown in *Figure 9.1*.

An alternative way of counting is to use an eyepiece graticule marked off in a grid. Although this will not allow a 'per ml' figure to be obtained, it is quite suitable for a quick comparison of different cultures.

9.2 Lengths and areas

For measuring lengths, eyepiece graticules are used [1]. The simplest form has a ruler marked out on the graticule which can be used to measure lengths in arbitrary units. To calibrate these measurements to

a.

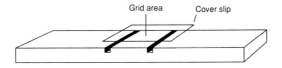

b.

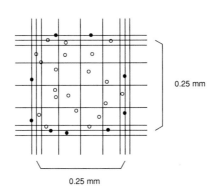

0.25 mm

Area = 1/16 mm^2
Depth = 0.2 mm

0.25 mm

Rules for counting:

- Do count all cells inside middle line
- Do count cells touching middle line on top and left sides
- Do not count cells touching middle line on bottom and right sides

e.g. If 20 cells are counted (open circles) inside the grid:
Volume under grid = 0.025 x 0.025 x 0.02 cm = 0.0125 cm^3
Cell density = 20 cells per 0.0125 ml = 1600 per ml

FIGURE 9.1: *Cell counting with a hemocytometer. **a.** Schematic diagram of a hemocytometer. **b.** Rules for counting and an example of a calculation.*

μm, a stage micrometer is used. This is a slide with a ruler marked off in absolute units (usually 100 μm in 1 or 2 μm divisions) and so by comparing the eyepiece units with the stage micrometer units the equivalent length of one eyepiece unit can be found in μm (*Figure 9.2b*). This calibration can then be used to measure the specimen (Figure 9.2c). Obviously the graticule will require recalibration for different objectives.

A more accurate device is a separate eyepiece that has within it movable cross-hairs controlled by a vernier screw (*Figure 9.2d*). These cross-hairs are superimposed on a ruler. They are lined up against one side of the specimen and the value is read off on the ruler and fractions on the vernier. The cross-hairs are then moved to the other side of the specimen, the procedure is repeated and the two readings subtracted to give the distance.

The device is calibrated to μm as before. Very accurate measurements can be performed with this (down to about 0.2 μm). There are all sorts of graticules for different purposes and suppliers are listed in Appendix D. Be careful to get the right diameter for your eyepieces.

Other measurements that can be made using the light microscope include measurement of area fractions and surface area:volume ratios as well as

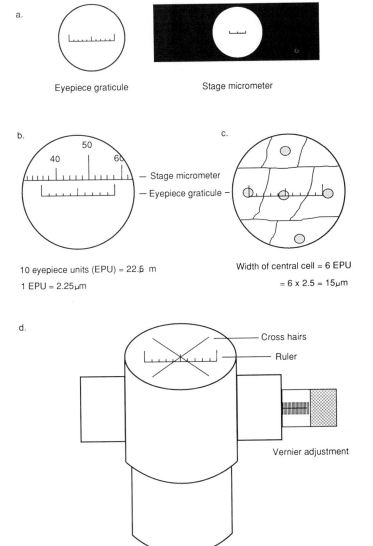

FIGURE 9.2: *Measuring lengths with an eyepiece graticule.* **a.** *Simple graticule mounted in a standard eyepiece and a stage micrometer.* **b.** *Eyepiece units calibrated with the stage graticule.* **c.** *Specimen measured using the eyepiece graticule and conversion to μm.* **d.** *Specialized eyepiece for more accurate measurement.*

shape, degree of curvature and others. These are performed using the technique of stereology, see references [2–4]. Stereology can either be done using a suitable eyepiece graticule or by overlaying photomicrographs with a transparent plastic sheet onto which a graticule has been copied. A commonly used graticule (one of the Weibel graticules [4]) is shown in *Figure 9.3*. This can be used to derive the surface area:volume ratio of a tissue from a sampled section. This parameter is important in the study of transport across the walls of, for example, microvilli in the intestine [5]. The first step is to measure the total length of all the lines on the graticule (L). Next superimpose the graticule over the image of the specimen and count the number of line ends (p) falling on the tissue of interest. Then count the number of times the lines intersect the boundary of the tissue (i). It can be shown that the surface area:volume ratio is given by:

$$\frac{S}{V} = \frac{4 \times p}{L \times i}.$$

An example of such a calculation is given in *Figure 9.3*.

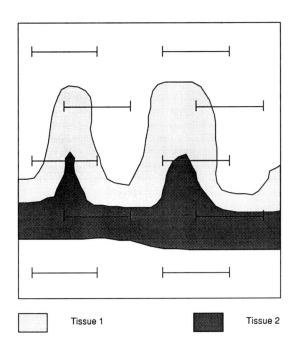

Tissue 1   Tissue 2

Total length of lines (L) = 10 x 5 = 50μm

Number of line ends in Tissue 1 (p) = 4

Number of intersections with Tissue 1 (i) = 8

Surface area:Volume = $\dfrac{4 \times 4}{50 \times 8}$ = 0.04

FIGURE 9.3: *Stereological estimation of surface area:volume using a Weibel graticule.*

It is important with all the measurement techniques to take repeated, random measurements over a specimen so as to get a representative sample. Methods for taking random samples are described in references [2–4].

References

1. Bradbury, S.J. (1991) *Basic Measurement Techniques for Light Microscopy,* Royal Microscopical Society Handbook No. 23. Oxford University Press, Oxford.

2. Bradbury, S.J. (1989) in *Light Microscopy in Biology: A Practical Approach* (A.J. Lacey, ed.). IRL Press, Oxford, p. 187.

3. Russ, J.C. (1986) *Practical Stereology.* Plenum Press, New York.

4. Weibel, E.R. (1979) *Stereological Methods, Vol. 1. Practical Methods for Biological Morphometry.* Academic Press, London.

5. Dunnil, M.S. and Whitehead, R. (1972) *J. Clin. Path.,* **25,** 243.

10 Photomicrography

10.1 Cameras and exposure systems

Photomicrography simply means taking photographs from a microscope and a basic knowledge of photography is assumed in this section. More details on the topics covered here can be obtained from references [1–4]. The equipment usually consists of a 35 mm camera body on top of a phototube with some sort of exposure meter attached. Modern microscopes often have several camera ports, allowing more than one camera body to be connected at the same time; this may include bodies of larger formats than 35 mm. The cameras are not quite the same as standard camera bodies as there is no mirror to be moved before exposure. Also, they do not require as fast a shutter speed as ordinary cameras; therefore the shutter system can be built more sturdily.

Most modern exposure meters use center-weighted average metering, which means that the meter calculates the average brightness over the whole field of view but biases it towards that in the center of the field. It then determines the exposure setting needed for the type of film being used. Obviously, therefore, the specimen should be put in the middle of the field of view. Taking a photograph is usually just a case of diverting the light from the eyepieces to the camera, setting the exposure time and pressing the exposure button. However, some images can be difficult to meter, especially if they are small bright spots on a dark background (e.g. fluorescence) or vice versa (e.g. dark field). Here, a spot metering system is better. This uses a much smaller area to calculate the exposure so with the object in this area one can be fairly confident that it will be exposed properly. Without a spot meter it is a case of trial and error with a range of exposure times to determine one that gives good results with your particular specimen. For example, with 800 ASA film (e.g. Kodak TMAX 400 film uprated to 800) an exposure time of 30 sec or more might be needed for fluorescently labeled cytoskeletal networks which cover the whole field (such as in *Figure 8.3e*). However, DAPI stained nuclei (such as in *Figure 8.3c*) often need only 1 sec or less, despite the meter indicating that a much longer exposure time is necessary.

10.2 Taking good photomicrographs

For an in-focus picture to be taken, the image at the camera must be par-focal (i.e. at the same focal plane) as the image at the eyepieces. The photo-tube is usually set up on installation so that further adjustment is unnec-essary. However, the eyepiece correction collars, which compensate for different people's eyesight, will still need adjusting. The procedure for this is described in Section 6.3.3. Always make sure that both the photoscreen *and* the image are in focus before taking a picture, in which case the photograph should be sharply focused.

When taking a photomicrograph, it is always advisable to consider how it might appear in an eventual publication. It is much easier to select from a series of negatives than to rephotograph a particular specimen. In addi-tion to the indicated exposure, take additional exposures at 2× and ½× this value ('bracketing the exposure'). Check that the orientation of the sample is optimal by rotating the stage or the camera. Check that the magnification is sufficient to fill the photographic frame while not too great to introduce empty magnification (see Section 10.4). Make sure that all of the desired portion of the specimen is within the photoframe. Con-sider if a colored filter to enhance the contrast of a stain would be useful.

Color can make all the difference to a presentation. *Figures 10.1–10.3* were taken using color slide film and the same specimens are presented in Chapter 8. In most cases, automatic exposure metering was used – either spot or average metering as appropriate. *Figure 10.1* shows the three different fluorochromes used to label an MRC 5 lung epithlelial cell; in this case the lens was a ×40 Plan–apo or Plan Neofluar oil immersion objec-tive. *Figure 10.1c* is a multiple exposure carried out by under-exposing each channel by half and not winding the film on between exposures. Fluorescence is ideal for this sort of picture because the colors are so attractive and, because the background is dark in each individual frame, there is still good contrast in a multiply-exposed image. *Figure 10.2* was taken with a ×63 Plan–apo oil immersion lens with the condenser oiled to the slide and Nomarski illumination. The film used was Kodak Ekta-chrome 200 daylight and so a pale gray/blue color balance filter was in-cluded to correct the color temperature of the tungsten lamp (see Section 10.3). The *Spirogyra* cell is optically sectioned inside the cell and the nucleus and nucleolus can be seen to be connected by strands to the cell wall. The spiral arrangement of chloroplasts is also clear. Finally, *Figure 10.3* is a dark field image of the *Antirrhinum* floral apex using a Didymium filter as well as a color balance filter. This adds a little pink to the reflected light areas. The cell walls are fluorescing blue as they are stained with Calcifluor. As this is dark field, the incident light does not reach the eyepieces (incidentally, before epi-illumination became universal, fluores-cence was always carried out this way).

10.3 Choice of film

For photomicrography, the main criteria are fine grain, medium contrast and for dim specimens, high sensitivity. For black and white photography, Kodak TMAX film is very popular as it has good contrast and low grain while being available at a range of different speeds. Much slower films (e.g. Agfapan APX 25 and Kodak Technical Pan) have a very fine grain and produce rather less contrast but give a very wide range of lights and darks (dynamic range).

What is 'speed' and why is one film 'slower' than another? Films are manufactured with different sensitivities to light. Sensitivity is described by the ISO (or ASA) number or 'speed' of a film. For example, a very sensitive film for use in low light conditions (e.g. fluorescence) will have a high ASA number (e.g. 800) and be called a 'fast' film. Agfapan APX 25 has a very low ASA number (25) and so is a 'slow' film. While this means that it is unsuitable for fluorescence, it is very good where a wide dynamic range is required. Increasing film speed normally gives increased graininess.

For some applications, especially photographing video monitors, ordinary processing of the negatives gives excessive contrast. The slow films mentioned above can be developed in such a way as to give much reduced contrast so the final negative will have full dynamic range. For example, to photograph a high resolution color video monitor, the following protocol gives good results: pictures are taken at 1 sec f 2.8 on Technical Pan film having set the contrast on the monitor to slightly less and the brightness slightly more than that which looks good to the eye. This gives an overexposed film which is then developed using a low contrast procedure with Kodak Technidol Liquid Developer to give a final negative with the right amount of contrast.

For color slides, film is designed for either tungsten or daylight color temperatures. Color temperature can be thought of simply as the amount of blue in a light. For instance, sunlight has a lot of blue in it whereas light from a domestic light bulb is much redder. Color slide films are designed to give the best color balance with either daylight or tungsten light. With the wrong type of film, a color cast will occur over the whole picture. For example, if a daylight film is used with a tungsten bulb, the pictures will be very yellow. Conversely, a tungsten film used with a UV bulb (which has a similar color temperature to daylight) will turn out rather blue. For color photomicrography with tungsten bulbs, as well as using the right color temperature rating film, the voltage on the bulb must also be set correctly. Varying the light intensity is commonly achieved by reducing the voltage as this is by far the most convenient. However, at low voltages, these bulbs give off a very red light. So for color photomicrography, it is

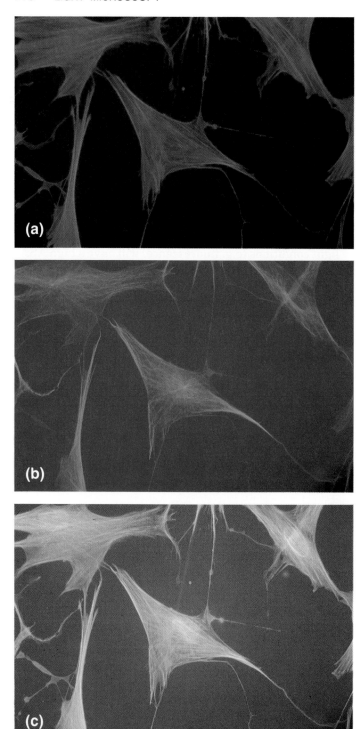

FIGURE 10.1: *MRC5 lung epithelial cells.* ***a.*** *Rhodamine (actin stress fibers).* ***b.*** *FITC (microtubules).* ***c.*** *Multiple exposure: rhodamine, FITC and DAPI.*

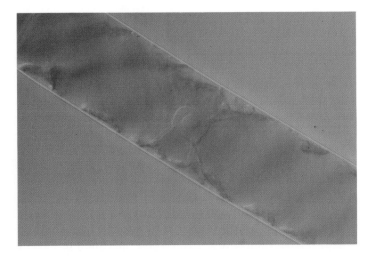

FIGURE 10.2: *Micrograph of* Spirogyra grevilleana *using Nomarski illumination.*

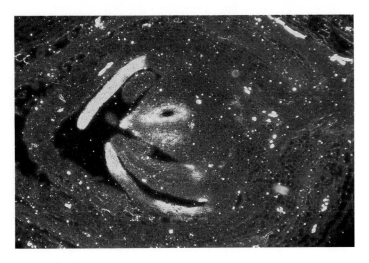

FIGURE 10.3: Antirrhinum *apex. Dark field.*

best to turn the lamp up to its optimum setting and reduce the light intensity using (gray) neutral density filters. The newer microscopes incorporate a device for automatic selection of the correct voltage. For color prints, there is no choice since there are no tungsten rated films available. A bluish color balance filter (e.g. from the Kodak Wratten range) can be used to correct the color temperature by cutting out some of the red from a tungsten bulb; otherwise yellowish pictures will result. An alternative solution is to take slides with tungsten rated film and then produce prints from the slides (e.g. Cibachrome prints).

10.4 Magnification

It is good practice to fill the frame of the photoscreen with your specimen; otherwise considerable enlargement will be needed when the film is printed, which can give rise to excessive graininess. Using the magnification changer (if fitted) can be useful here to give intermediate magnifications between objectives but can reduce the light intensity significantly. Filling the frame may not be possible, however, because of problems with depth of field. The depth of field of an objective is the distance either side (up and down) of the in-focus image that is still reasonably in focus. This decreases with increasing numerical aperture so, for example, a ×63/1.4 NA objective will have a very much narrower depth of field than a ×10/0.4 NA objective. Therefore, to get the whole specimen in focus a lower magnification objective may be required than that which will fill the frame. An alternative approach is optical sectioning, described in Sections 5.1 and 7.8.3.

Useful magnification (see Section 2.2) for the eye is about 1000×NA for well contrasted points. With film, this is reduced to about 200×NA total magnification at the film because of its grain. As long as the best resolution is recorded on film, the limit to the final print magnification is about the same as by eye, that is, 1000×NA. This is for a person with perfect eyesight, however, and so it may be useful to increase this slightly to make the fine detail easier to see. Magnification for video microscopy is discussed in Section 11.4. For slide film, the 200×NA limit on the film should be adopted.

This leads on to the actual magnification at the camera, which depends on the arrangement of the phototube and projector lens as well as the objective. Never rely on data given in the handbook; always measure it for yourself (it is often not the same as objective magnification×eyepiece magnification). Place a stage micrometer (see Section 9.2) under the objective and take some photographs using different settings of the magnification changer if fitted. When the film is developed, calculate the magnification from the negative. To get a quick estimate of the approximate magnification at the film, simply open the camera back and project the image of the stage micrometer onto a piece of lens tissue, placed where the film normally lies, and measure its length with a ruler.

10.5 Printing

A full description of photographic printing requires a book in itself but there are some points about black and white printing that are particularly relevant to microscopy. First, unless all the pictures on a film are taken in exactly the same way, the range of contrasts between different negatives can be very great. Multigrade papers offer a great advantage in this respect.

Fluorescence pictures pose particular problems that need to be addressed carefully. Often a dark background is required, while at the same time all the detail in the bright structures must be visualized. It might seem logical to use a high contrast paper to get this result but this is exactly what not to do. This will certainly give a dark background, but all the bright areas will be brilliant white with very little dynamic range. Instead, use a low contrast paper and experiment with the exposure until the bright areas are just beginning to lose their whiteness. This will give the best range of grays over the structure and a reasonably black background.

It is especially important with microscopy to know the magnification of the final print. Having calibrated the microscope (see Section 10.4) and with the negative in the enlarger, measure the smallest side of the picture frame when the image of your specimen is the desired size. For 35 mm film, divide this by 24 mm (the height of the picture frame on the negative) to get the magnification at the print. Magnification can be quoted in these terms (e.g. ×1200) or by calculating the size of a standard unit scale bar (e.g. 5 μm, 10 μm, 100 μm, etc.) at this magnification and using Letraline or something similar to put a scale bar on the actual print. The bar is then defined in the figure legend: 'bar = 5 μm', for example. This method is preferable as it allows the reader easily to assess the size of structures in the picture and to compare results between different workers.

Color printing is more difficult than black and white and most workers have their color negatives printed professionally. Even high-street photoprocessors can produce very respectable prints without knowing what the final image is *supposed* to look like.

10.6 Stereo pairs for three-dimensional images

Stereoscopy, as taking and viewing stereo pairs is called, is a fascinating pastime for both the amateur and the professional. In order to see the stereo effect in stereo pairs, either a stereo viewer can be used or else fusing the images can be achieved by eye – an ability which some people find difficult to acquire. For printing stereo pairs, unless it is known that the reader will have access to a stereo viewer, the main criterion is to print the two images so that they can be easily fused. For this, the distance between equivalent points on both images should be close to the average distance between the eyes. This has led to a standard distance apart of equivalent points of 6.5 cm. The result of this is that the pair cannot be more than 13 cm across; a size which fits most journals. More details on stereoscopy are given in reference [4].

To fuse stereo pairs without a stereo viewer, hold the picture quite close up and relax your eyes until you can see three images: the separate projections at the side and the fused one in the middle. This middle one will usually be out of focus at first but with a lot of patience and saying to yourself 'relax' and 'merge' at the same time, the fused image will gradually come into focus. It is very rewarding when you are able to get a clear, fused image in this way. However, fusing the stereo pair like this reverses the stereo depth, i.e. the back of the specimen appears at the front and *vice versa*. This is not normally a problem but to visualize the stereo image the right way round you have to *diverge* your eyes rather then converge them as before. This is much more difficult for most people and it is easier to get a stereo viewer.

References

1. Delly, J.G. (1988) *Photography Through the Microscope*. Eastman Kodak Co., Rochester, NY.

2. Thomson, D.J. and Bradbury, S.J. (1987) *An Introduction to Photomicrography,* Royal Microscopical Society Handbook No. 13. Oxford University Press, Oxford.

3. Evenett, P.J. (1989) in *Light Microscopy in Biology: A Practical Approach* (A.J. Lacey, ed.). IRL Press, Oxford, p. 61.

4. Turner J.N. (1981) in *Three Dimensional Ultrastructure in Biology* (J.N. Turner, ed.) Academic Press, Oxford, p. 1.

11 Video Microscopy

There has been a tremendous expansion in the use of video in microscopy in recent years. This is probably not due to the most obvious advantage of video, that of real-time imaging of moving events, but to its use in image analysis systems. The cheapness of PCs and the associated hardware and software to store and manipulate images, coupled with the availability of low-cost CCD cameras has made the analysis of microscope images commonplace. In this chapter the equipment used and the sorts of analysis which can be performed will be briefly outlined. The standard text on video microscopy is that of Inoué [1]; a more recent review is by Weiss *et al.* [2] while the major video equipment manufacturers will also provide literature and advice on the more up-to-date equipment.

11.1 Types of video camera

Video cameras using pick-up tubes work by detecting electrical charge built up on a photoreactive plate when light falls on it. The light first passes through a positively charged, transparent window and then hits a deposited layer of a photoreactive substance. This changes its conductance, depending on the amount of light that falls on it, and causes charge to be drawn from the charged window and stored at the back of the layer. An electron beam is then scanned over the back of the layer and interacts with the stored charge, producing a potential difference which gives rise to the current that makes the picture. The difference between the different types of video camera is basically the substance used to make the photoreactive layer. These give different sensitivities for different applications, for example, infra-red, UV, low-level visible light etc. The most commonly used cameras for transmitted light microscopy are Vidicon and Chalnicon, the latter being more sensitive. For low light levels, for example, fluorescence, SIT (silicon intensifier target) cameras are often used. These usually come with external control electronics to allow control of brightness and contrast (called gain and black level). The most sensitive video camera is the ISIT (which stands for intensified SIT).

The phrase video camera now also encompasses cameras based on a type of silicon chip called a charge-coupled device (CCD). This can be thought of as a block containing an array of very small wells. When a photon of light enters a well, an electron is stored there and these electrons build up all the time the camera shutter is open. When the shutter closes, the number of electrons in each well determines the brightness of the corresponding picture element (pixel) which make up the image when it is displayed. One of the main advantages of CCD cameras is that because there is no electron gun, they are lighter and more cost-effective. Using a slow scan (say 1 image per second), rather than at video rates, the sensitivity of the best CCD cameras is such that single photons can be detected.

Most cameras used for video microscopy are monochrome, although the home video market has meant that color video cameras (which are all CCD based) are now affordable and useful for applications where light levels are reasonably high.

11.2 Fitting a video camera to a microscope

Most video camera systems come ready fitted with appropriate connectors and adaptors to fit onto your microscope if it already has a 35 mm camera setup. If not, the main problem is to project the image through the camera window in such a way that the image is focused on and fills the pick-up tube or CCD chip. This can be done either with or without another lens. As a first step, remove the cine-lens from the camera (if there is one) and support the camera 20–30 cm from the eyepiece of the microscope. Shield the gap between the two with black paper or card to keep out stray light and adjust the gap between the two until a focused image is produced. Most commercial systems incorporate a low power ($\times 1$ to $\times 4$) eyepiece or projector lens as for photomicrography. If using a photomicroscope, the phototube will need to have a male 'C-mount' adaptor at the top to screw into the camera. This is a fairly uniform standard for video cameras but not camcorders. Another standard is the size of the target, which is normally 1 inch.

Tube cameras as opposed to CCD cameras can be ruined by focusing too much light onto the target. Therefore, with whatever projection system is used, make sure the lamp is turned right down before diverting the light to the camera.

The simplest video system is a camera, a black and white monitor (not a television unless it has a video input) and a length of coaxial cable with a suitable connector on each end.

11.3 Image capture, display, processing and analysis

Video images can be 'captured', that is, converted to a stream of digital information and stored as files on a computer using a frame-store (also called a frame-grabber). These are often PC based and are simply an extra card for a PC with connections for the camera and monitor. There are now many manufacturers of frame-stores either as stand-alone units or as a part of image analysis packages. All frame-stores have the facility to display the images stored and often to adjust the brightness and contrast too. This is usually but not always on a separate monitor from the computer.

Video images can be displayed on a black and white monitor, or, depending on the frame-store on a color monitor using either 'composite' video or separate channels for red, green and blue (RGB). The size of screen is not as important as the number of dots (pixels) that make up the image.

Image processing means altering the stored image, or in real time on a live image. At its simplest this means altering the brightness and contrast. In addition there are programs for improving noisy images, sharpening the edges of objects within an image, cutting and pasting parts of an image, merging one image with another, rotation and mirror imaging, negative contrast, producing diffraction patterns, 3D reconstruction of optical sections, making pseudo-colored images and many more. More details on image processing can be found in references [1–4] or by contacting the companies listed in Appendix D.

Image analysis is subtly different from image processing but the two techniques are closely related and the terms often interchanged. Analysis of an image could include plotting the intensities of all the pixels within a defined area or along a line (e.g. *Figure 11.1*), measuring parameters of an area of the image traced with a mouse (e.g. perimeter, area, average pixel intensity with standard deviation), counting the number of a particular object, perhaps defined by 'thresholding' at a particular intensity (an image analysis function giving a particular color to all pixels within a specified intensity range) and particle analysis.

11.4 Magnification, pixel size and resolution

In Section 10.4 magnification for photomicrography was considered. The same principles apply here except that the video monitor has considerably less resolution than photographic film. Typically, an image will be

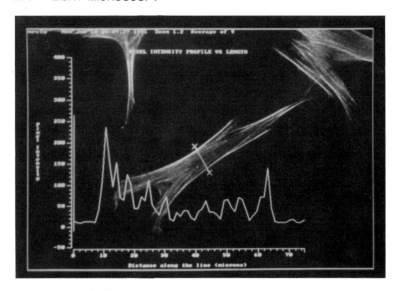

FIGURE 11.1: *MRC5 cell using confocal imaging of the rhodamine fluorescence. Plot of the pixel intensity (y-axis) along a line defined by the operator (x-axis). This gives an indication of the spacing of the stress fiber bundles.*

displayed on a 512×512 array of pixels. The resolution of the light microscope itself is about 0.2 μm in the x and y planes so that is the minimum distance that can be represented on two adjacent pixels without producing empty magnification. This means that the minimum width that a whole screen can represent is 512×0.2 = 102.4 μm. Therefore, there is no advantage in the camera and projection system producing an image much smaller (i.e. a much greater magnification) than this as no more resolution will be achieved. This corresponds to a magnification of about ×40 for the objective and projector lens combined. However, the optical system of the microscope (effectively the objective NA) has to be capable of the best resolution resolving two points that far apart, and as this requires a high NA lens (e.g. ×63/1.4), in practice it is best to use the highest NA lens possible and to ignore any empty magnification produced. This strategy will also ensure the brightest image possible. Of course, a larger image may also be more pleasing to look at.

11.5 Recording dynamic processes

It is straightforward to connect a video camera to a domestic VHS video recorder for recording moving events. However, the quality of the image produced will be better with one of the professional formats (e.g. Super VHS or U-matic). It is also worth considering a machine that will produce and record text and possibly simple graphics such as arrows on the screen.

For long term dynamic events, a VHS time-lapse video recorder is used. This records a single video frame at intervals up to 8 sec or more and also moves the tape much more slowly than normal. In this way, 1000 h of action can be reduced to 3 h of tape played back at the normal speed. Most time-lapse recorders have a text-recording facility.

Aesthetic considerations are particularly important if video-recorded material is to be published because as yet, the only practical way to do it is to take photographs. The resolution of VHS is much less even than a standard monitor and so it is important to frame the shot and adjust the brightness and contrast carefully.

The major alternative to video-recorders are laserdisks which can store many hundreds of 512×512 images on a single disk and retrieve them within a second or less without loss of resolution. This rapidly developing area of technology is likely to produce real-time playback soon.

References

1. Inoué, S. (1986) *Video Microscopy*. Plenum Press, New York.

2. Weiss, D.G., Maile, W. and Wick, R.A. (1989) in *Light Microscopy in Biology: A Practical Approach* (A.J. Lacey, ed.). IRL Press, Oxford, p. 221.

3. Castleman, K.R. (1979) *Digital Image Processing*. Prentice-Hall, London.

4. Kennedy, J.M.J. (1991) *Micros. Anal.*, **26**, 5.

Appendix A.
Further Reading

The following texts will be useful for those readers who wish to investigate light microscopy further.

Lacey, A.J. (1989) *Light Microscopy in Biology: A Practical Approach.* IRL Press, Oxford.

James, J. and Tanke, H.J. (1991) *Biomedical Light Microscopy.* Kuher Academic, Dordrecht.

Bradbury, S. (1989) *An Introduction to the Light Microscope.* Royal Microscopical Society Handbook No. 1. Oxford University Press, Oxford.

Spencer, M. (1982) *Fundamentals of Light Microscopy.* Cambridge University Press, Cambridge.

O'Brien, T.P. and McCully, M.E. (1981) *The Study of Plant Structure. Principles and Selected Methods.* Termacarphi Pty Ltd, Australia.

Taylor, D.L. and Wang, Y.-L. (1989) *Fluorescence Microscopy of Living Cells in Culture, Vols A and B.* Academic Press, London.

Microscopy and Analysis. Rolston Gordon Communications. A trade magazine with useful short articles on a wide range of microscopical topics. Also has advertisements from microscopical manufacturers. Published bimonthly.

Journal of Microscopy. Royal Microscopical Society and Blackwell Scientific Publications. The authoritative UK journal on microscopy. Published monthly.

Appendix B.

Glossary

Abbe condenser: a simple condenser designed by Ernst Abbe (1840–1905). Condensers of this type are corrected for spherical but not chromatic abberation.

Accessory lens: the name for the lens used to broaden the area of illumination when viewing at low magnification.

Amplitude: the maximum oscillation from zero that a wave achieves. The amplitude of a light wave determines the brightness of the light.

Analyzer: a polar normally placed above the specimen and crossed with respect to the polarizer. Only light that has had its plane of polarization altered will pass through the analyzer hence it 'analyzes' this polarization change.

Apochromatic: a property of a lens which has been corrected for chromatic aberration.

Autofluorescence: the inherent ability of a specimen to fluoresce.

Back focal plane: the focal plane of a lens which lies behind the lens when looking in the direction of the passage of light.

Beam splitter: a prism or prisms that divert part or all of the light from one direction to another. A beam splitter is often used to divert some or all of the light from the eyepieces to the camera.

Bertrand lens: a lens that allows the back focal plane of the objective to be viewed through the eyepieces. It works by transferring the image of the back focal plane to the primary image plane.

Birefringence: a property of a transparent specimen with two different refractive indices in different orientations to split plane polarized light into two components.

Bright field: transmitted light illuminates the specimen directly, allowing it to be seen.

Center-weighted average: a form of camera exposure metering that gives priority to objects in the center of the field of view.

Chalnicon: a type of video camera with medium sensitivity suitable for transmitted light microscopy.

Chromatic aberration: a property of a lens to magnify different wavelengths of light to different extents.

Collector lens: a lens used with a light source to provide a suitably sized area of illumination.

Color temperature: the temperature (in degrees Kelvin) of an ideal black body emitting radiation of the same color quality as that from the light source being described.

Compensator: an optical element that retards light to a variable, known amount; used in polarized light microscopy to determine the physical structure of a birefringent specimen.

Compound microscope: produces magnification using an objective and an eyepiece.

Condenser: part of a microscope designed to collect, focus and project light onto the specimen.

Condenser aperture: a variable aperture in the condenser used in Köhler illumination to vary the numerical aperture of the illumination.

Confocal aperture: an aperture placed on an equivalent point on the light path to the focal plane of the objective. It is used to reject out-of-focus light producing in-focus optical sections.

Confocal laser scanning microscope (CLSM): uses a scanning system consisting of rotating mirrors or opto-acoustic deflectors to scan a point of laser light over the specimen.

Confocal microscopy: uses a confocal aperture to produce in-focus optical sections.

Convex (converging) lens: a lens which is thicker in the middle than at the edges and can form a real image.

Correction collar: an adjustable ring on an objective. There are two types, one controlling a diaphragm in the back focal plane which allows the NA to be adjusted and the other allowing compensation for different immersion media.

Cryo-fixation: rapid cooling of a specimen to preserve its structure.

Cryostat: a trade name for a refrigerated microtome used for cutting frozen sections.

Cytospin: a trade name for a centrifuge used for preparing spread cell preparations on slides.

Dark field: oblique illumination is used to produce a dark background with reflective structures brightly contrasted against it.

Deblurring: a mathematical process used to remove out-of-focus blur from digitized optical sections.

Depth of field: the distance in the z axis either side of the specimen that is in-focus.

Dichroic mirror: a mirror made by depositing a fine metal layer onto glass which has the property that it reflects light at one range of wavelengths and transmits light at other wavelengths. They can be made with one, two or three reflection ranges (single, double and triple dichroics).

Diffraction: a change in the direction of light caused by passing through a substance with a higher refractive index than air (> 1.0).

Dynamic range: in the context of photomicrography is the number of different shades of gray between black and white that are present in the image.

Emission filter: a filter (often made of colored glass) to ensure that only light from the fluorochrome gets to the eyepieces or camera.

Empty magnification: greater magnification than the useful magnification. No useful information will be obtained and sharpness and contrast will decrease.

Epi-illumination: light passes through the objective before reaching the specimen. The objective therefore also acts as a condenser. Fluorescence microscopy is normally performed with epi-illumination.

Excitation filter: a filter (usually of the interference type) which limits the incident light to specific wavelengths (i.e. those that excite the fluorochrome).

Exit pupil: the point just above an eyepiece where an image is formed that can be viewed by placing the eye at that point.

Eyepiece: magnifies the image formed by the objective allowing it to be seen by the eye.

Eyepiece graticule: a transparent disk marked with a scale or other measuring device that fits into the eyepiece and is superimposed on the image of the specimen.

Field aperture: a variable aperture below the condenser (often on the stand) used to vary the area of the slide that is illuminated and to set up Köhler illumination.

Filter holder: a holder for colored or polarizing filters usually placed below the condenser so that it can be swung into the light path as required.

Fluorescence: the property of a molecule to emit light at a specific range of wavelengths when hit by incident light of a shorter wavelength.

Fluorochrome: a molecule that exhibits fluorescence. Many are now available conjugated to antibodies, etc.

Focal point: the point on one side of the optical axis of a lens where parallel light entering the lens from the other side is focused.

Frame-store: an electronic device to allow images from video cameras, etc., to be digitized and stored as computer memory. Frame-stores often also have image processing capabilities.

Gain: the electronic control of brightness on a video image. An offset control (or black level) is used in conjunction with the gain control to produce an image with the best contrast.

Graticule: any pattern superimposed over an image. Can be printed onto a transparent sheet overlaying a photomicrograph or onto a transparent disk to fit into the eyepiece.

Hemocytometer: a device for counting cells. It consists of a glass slide and coverslip forming a counting chamber of known volume and is marked off in a grid to aid cell counting.

High eyepoint eyepiece: an eyepiece with its exit pupil further from the top surface of the eyepiece than normal to accommodate people who wear spectacles.

Image analysis: performing measurements on features in a stored image without changing the image itself.

Image processing: altering the image to improve contrast, sharpness, etc.

Infinity-corrected objectives: have a tube-length of infinity and so require a separate lens to form an image. Their advantage is that the light from the image is kept as parallel rays until just before the eyepieces which means that optical elements can be introduced anywhere along the light path without affecting the focus or magnification of the final image.

Interference: the interaction between light waves.

Inverted microscope: used for examining specimens in Petri dishes, etc. The condenser is mounted above the stage with the objectives underneath.

Köhler, August: (1866–1948) a German biologist who developed the bright field illumination system universally used nowadays for transmitted light microscopy.

Magnification: the (apparent) increase in size of an object achieved in the image of the object using lenses.

Magnification changer: a set of lenses that increase the magnification of the image.

Monitor: a TV tube designed to take video or RGB inputs.

Nipkow disk: a disk with a spiral pattern of holes arranged so that as it spins, light shining through the disk scans every part of the specimen with a series of small points of light. It is used in the tandem scanning method of confocal microscopy.

Nomarski differential interference contrast (DIC): an illumination technique using polarized light that produces an apparent 3D effect by creating light and dark shadows at opposing edges of features in the specimen. (The right to use the name 'Nomarski' is held by Zeiss.)

Numerical aperture (NA): can be thought of as the light gathering power of a lens. NA depends on the angular aperture of the lens and the refractive index of the medium between the lens and the specimen.

Objective: the essential part of any microscope – it produces the most magnification and its numerical aperture limits the resolution achievable in the image of the specimen.

Oiling the condenser: placing immersion oil between the condenser and the underneath of the slide.

Optical sectioning: the process of recording images at different focus planes through the specimen using a high NA objective. The images can be enhanced to remove out-of-focus blur and then reconstructed into a 3D image. *See also* confocal microscopy.

Optovar: another word for a magnification changer. Some also contain the Bertrand lens.

Parfocal: is a property of objectives such that two objectives that are parfocal with one another can be interchanged without having to alter the focus controls.

Patch stop: is an opaque disk that is used in the condenser to provide oblique illumination for dark field microscopy.

Phase annulus: a ring shaped aperture placed in the condenser to produce illumination for phase contrast microscopy. A particular phase annulus will have a matching sized ring in the phase plate in the objective.

Phase contrast: uses the retardation of light by the specimen to produce phase differences which are converted into contrast.

Phase difference: between two waves is the number of wavelengths which one wave lags behind the other.

Phase plate: the glass plate placed at the back focal plane of a phase objective which contains the phase ring.

Phase ring: a darkened ring on the phase plate which (normally) retards light less than the phase plate. It is used in phase contrast microscopy together with the phase annulus to produce contrast from phase differences between light rays from different parts of the specimen and the background.

Phase telescope: a modified eyepiece that can be focused in the same way as a Bertrand lens to view intermediate image planes.

Photomicrography: taking photographs down a microscope.

Photomultiplier tube: works like a video camera but operates at a slower rate and so is usually more sensitive. Photomultiplier tubes are used in confocal microscopes to detect the returning scanned fluorescence.

Photoscreen: a transparent plate overlaid with a frame (photoframe) corresponding in size to the area of the microscope image that will be recorded on film by the camera.

Pick-up tube: the part of a video camera that contains the photoreactive plate.

Pixel: a picture element. Arrays of pixels make up an image on a video or on a monitor.

Plan: or planar, refers to a lens that has been corrected so that the resultant image is in-focus across the whole field of view.

Plane polarized light: light oscillating in one plane only.

Polar: anything that transmits light in one plane only. They are often made of a stretched plastic film sandwiched between two layers of glass. These polars transmit light oscillating in a plane parallel to the stretch.

Polarizer: a polar placed in the incident light (usually below the condenser) to produce plane polarized light.

Primary image plane: the image plane where an image of the specimen is first formed.

Projection: the process of calculating what a set of serial or optical sections would look like if stacked on top of one another and photographed from different angles.

Quarter wave plate: retards light by a quarter of a wavelength and produces circularly polarized light.

Real image: an image which can be seen on a screen

Reflected light microscopy: uses plane polarized (epi-polarization microscopy) or circularly polarized (reflection contrast microscopy) light to produce contrast from reflective surfaces.

Refractive index: the ratio of the speed of propagation of light through a vacuum to that through the specimen.

Resolution: the ability of an optical system to distinguish fine detail in a specimen.

Semi-silvered mirror: one that reflects half the incident light and transmits the rest.

Serial sections: adjacent physical or optical sections completely sampling the tissue.

Stage micrometer: a scale marked in absolute units (e.g. μm) used to calibrate eyepiece graticules, cameras, etc.

Stereo microscope: one that effectively is two monocular microscopes mounted side-by-side allowing stereo depth perception.

Stereology: the technique of estimating areas, volumes and other parameters of the whole tissue from sections.

Stereoscopy: making and viewing stereo images to produce stereo depth.

Tally counter: a device to help in counting large numbers of, for example, blood cells using a hemocytometer.

Tandem scanning microscope (TSM): uses a Nipkow disk to produce confocal images using non-laser illumination.

Thresholding: setting all the pixels in a certain range of intensity values to a particular value or color.

Tomography: collecting serial sections and reconstructing them to form a 3D image.

Transmitted light: light passes through the specimen before reaching the objective.

Tube length: the physical distance between the objective and the eyepiece.

Useful magnification: the maximum magnification that will give useful information from the specimen. Usually regarded as $1000 \times NA$.

Vibratome: a trade name for a vibrating microtome.

Vidicon: a type of sensitive video camera suitable for fluorescent microscopy.

Virtual image: an image which cannot be seen on a screen but which can be converted to a real image by a converging lens.

Wavelength: the distance between two equivalent points on a wave (e.g. peak to peak).

Working distance: the distance from the front lens of the objective to the cover slip.

Appendix C.

Fluorescence Filter Sets

Filters for fluorescence microscopy can be grouped into three categories: band pass, long (or short) pass and dichroic mirrors. Examples of all three are shown in *Figure 4.8*. Band pass filters block light at two wavelengths (cutoffs) close together and so transmit light of one color only. For instance, the DAPI excitation filter transmits UV light only. Band pass filters are either specified by their peak transmission wavelength and the bandwidth between 50% transmission wavelengths (in this case BP 365/12) or just as the 50% transmission wavelengths (e.g. BP 450–490 for blue excitation). Dichroic mirrors reflect light (i.e. no transmission) between a fairly distinct range of wavelengths and transmit at other wavelengths. They are usually specified by the wavelength at 50% transmission on the long side. In the DAPI example, the Zeiss dichroic mirror is called FT 425. Long pass filters simply transmit light longer than a certain wavelength. Again, the specification is the 50% transmission wavelength (e.g. LP 450 for DAPI emission). The letters 'SP' (which stand for short pass) may also be seen; these filters transmit light shorter than a certain wavelength. Long and short pass filters are often made of colored glass but higher specification filters (i.e. with sharper cutoffs and better transmission and rejection of light), band pass filters and dichroic mirrors are made by coating glass with extremely thin layers of metal. These are called interference filters as they work by using destructive interference (see Section 4.2) to block the light at certain wavelengths.

Listed below are the components of the fluorescent filter sets specifically for DAPI, FITC and TRITC (see Section 4.3) supplied by the four main microscope manufacturers. The information was taken from the manufacturers' literature and uses their names for the filters but the exact details should be checked with them before ordering. Also indicated are some other commonly used fluorochromes which can be used with these filter sets, although there may be a set specially designed specifically for that fluorochrome. There are also many other fluorochromes available that are not listed. Again, the manufacturer will help. If you know that the excitation and emission wavelengths of a particular fluorochrome are approximately the same as the ones below, it is worth trying these sets first before buying a specialized set.

DAPI (UV light). These filter sets will also work with Calcifluor white, AMCA and Hoechst 33258 and 33342

Manufacturer	Filter set	Excitation filter	Dichroic mirror	Emission filter
Leitz	A	BP 340-380	RKP 400	LP 430[a]
Nikon	UV-1A	EX 365/10	DM 400	BA 400[a]
Olympus	BH2[b]-IV	UG-1	DM 400	L 420[a]
Zeiss	UV-H 365	BP 365/12	FT 395	LP 397[a]

[a]A specific barrier filter (e.g. BP 450–490) used instead of a long pass filter will make the signal more blue looking.
[b]BH2 is a particular microscope.

FITC (blue light). These filter sets will also work for acridine orange, ethidium bromide, propidium iodide, quinacrine and chromomycin

Manufacturer	Filter set	Excitation filter	Dichroic mirror	Emission filter
Leitz	I3	BP 450-490	RKP 510	LP 515[a]
Nikon	B–2A	EX 450-490	DM 510	BA 520[a]
Olympus	BH2-B	BP 490	DM 500	BP 520[a]
Zeiss	Blue 450-490 IFB	BP 450-490	FT 510	LP 515[a]

[a]A specific barrier filter (e.g. BP 520–560) used here instead of a long pass or less specific barrier filter will cut out any red light. Alternatively, this barrier filter can be kept in a subsidiary filter holder and inserted when required.

Rhodamine (usually TRITC but also XRITC; green light). These filter sets will also work with Texas Red and to some extent phycoerythrin

Manufacturer	Filter set	Excitation filter	Dichroic mirror	Emission filter
Leitz	L2.1	BP 515–560	RKP 580	LP 580
Nikon	G-2A	EX 510–560	DM 580	LP 590
Olympus	BH2-G	BP 545	DM 570	BP 545
Zeiss	Green 510–560	LP 510	FT 580	LP 590

Appendix D.

Suppliers

Below is a list of UK suppliers for the consumables, accessories and equipment mentioned in the text. Addresses and telephone numbers are given at the end.

Anti-fade mounting media (glycerol based): 'Citifluor' containing DABCO available from Citifluor Ltd. Propyl gallate (available as a solid from Sigma): make a 5 (w/v) solution in buffered 90% (v/v) glycerol, mix gently overnight at room temperature and store at 4°C.

Confocal microscope systems: At the time of writing, the following systems were available: CLSMs: BioRad MRC600 distributed by BioRad Microscience; Leica CLSM distributed by Leica UK Ltd; Meridian Instruments Insight distributed by Biotech Instruments; Noran Instruments Odyssey distributed by Tracor Europa; Sarastro 2000 CLSM distributed by Molecular Dynamics Ltd; Zeiss CLSM distributed by Carl Zeiss (Oberkochen) Ltd. TSMs: Noran Instruments TSM distributed by Tracor Europa; Syncotec K2S-Bio distributed by Syncotec, Germany.

DePeX and Euparal: 'Gurr' microscopy materials available from BDH (Merck), Agar Scientific and Taab.

Dustoffs: Most photographic shops and suppliers. The compressed gas ones are easiest and now are all ozone-friendly.

Eyepiece graticules, stage micrometers: Graticules Ltd., Agar Scientific, Taab and some of the general scientific suppliers above.

Fluorescent probes and molecules: Molecular Probes, distributed in the UK by Cambridge BioScience.

Filters for epi-fluorescence microscopy; polars and other filters: A wide range of optical filters is made by Omega Optical Inc. in the US. They are distributed in the UK by Glen Spectra.

Haemocytometers and tally counters: Any scientific supplies firm.

Image processing: Systems and software. Synoptics, Seescan and Ai Cambridge are major companies but many others advertise in the trade magazines (e.g. *Microscopy and Analysis*).

Immersion oils: with different refractive indices. Cargille laser liquids.

Lens tissue: Whatman 105. Available from Agar Scientific and other general scientific suppliers (see below).

Microscopes: The main manufacturers are Olympus, Nikon, Carl Zeiss

(Oberkochen), Leitz (sold by Leica UK) and BioRad Nanoquest Ltd (the M17 range which derive from Vickers). General scientific suppliers often sell microscopes for routine work and some sell ones for specialist applications, e.g. materials or medical. Others are advertised in trade magazines such as *Microscopy and Analysis* (see Appendix A).

Microscopical stains, e.g. toluidine blue: 'Gurr' microscopy materials available from BDH.

Slides, cover slips, staining equipment, general microscopical supplies: Agar Scientific and Taab are microscope specialists. All the general scientific suppliers also have some microscopical accessories in their catalogues. The following have a better than average range: BDH, Hawfell Ltd, Fisons and Bio-Rad Microscience.

Stereoscopy accessories: Agar Scientific.

Addresses

Ai Cambridge Ltd, London Road, Pampisford, Cambridge CB2 4EF, UK. Tel: 0223 834420; Fax: 0223 835050.

Agar Scientific Ltd, 66a Cambridge Road, Stanstead, Essex CM24 8DA, UK. Tel: 0279 813519; Fax: 0279 815106.

BDH Merck Ltd, Customer Service Center, Hunter Boulevard, Magna Park, Lutterworth, Leics. LE17 4XN, UK. Tel: 0800 223344; Fax: 0455 558586.

BioRad Microscience Ltd, BioRad House, Maylands Avenue, Hemel Hempstead, Herts. HP2 7TD, UK. Tel: 0442 232552; Fax: 0442 234434.

BioRad Nanoquest Ltd, Haxby Road, York, N. Yorks. YO3 7SD, UK. Tel: 0904 645624; Fax: 0904 645624.

Biotech Instruments Ltd, 183 Camford Way, Luton, Beds. LU3 3AN, UK. Tel: 0582 502388; Fax: 0582 597091.

Cambridge BioScience, 25 Signet Court, Stourbridge Common Business Centre, Swann's Road, Cambridge CB5 8LA, UK. Tel: 0223 316855; Fax: 0223 60732.

Cargille (RP) Laboratories Inc., 55 Commerce Road, Cedar Grove, New Jersey, USA.

Citifluor Ltd, The City University, Northampton Square, London EC1V 0MB, UK. Tel: 071 253 4399, ext. 2502.

Fisons Scientific Equipment, Bishop Meadow Road, Loughborough, Leics. LE11 0RG, UK. Tel: 0509 231166; Fax: 0509 231893.

Glen Spectra, 2–4 Wigton Gardens, Stanmore, Middx HA7 1BG, UK. Tel: 081 204 9517; Fax: 081 204 5189.

Graticules Ltd, Morley Road, Tonbridge TN9 1RN, UK. Tel: 0732 359061; Fax: 0732 770217.

Hawfell Ltd, Hoverton Street, Cambridge, UK. Tel: 0223 212323.

Leica UK Ltd, Davy Avenue, Knowhill, Milton Keynes MK5 8LB, UK. Tel: 0908 666663; Fax: 0908 609992/3.

Molecular Dynamics Ltd, 4 Chaucer Business Park, Kemsing, Sevenoaks, Kent TN15 6PL, UK. Tel: 0732 62565; Fax: 0732 63422.

Nikon UK Ltd, Haybrook, Halesfield 9, Telford, Shrops. TF7 4EW, UK. Tel: 0952 587444; Fax: 0952 588009.

Olympus Optical Co. (UK) Ltd, 2–8 Honduras Street, London EC1Y 0TX, UK. Tel: 071 253 2772.

Seescan plc, Unit 9, 25 Gwydir Street, Cambridge CB1 2LG, UK.

Sigma, Fancy Road, Poole, Dorset BH17 7NH, UK. Tel: 0800 272572; Fax: 0202 715460.

Syncotec, Loherstraße 4, D-6334, Aßlar, Germany.

Synoptics Ltd, 271 Cambridge Science Park, Milton Road, Cambridge CB4 4WE, UK. Tel: 0223 423223; Fax: 0223 420020.

Taab Laboratories Equipment Ltd, 3 Minerva House, Calleva Industrial Park, Aldermaston, Reading, Berks. RG7 4QW, UK. Tel: 0734 817775; Fax: 0734 817881.

Carl Zeiss (Oberkochen) Ltd, P.O. Box 78, Woodfield Road, Welwyn Garden City, Herts. AL7 1LU, UK. Tel: 0707 331144; Fax: 0707 330237.

Index